Suzy Koontz

MULTIPLY WITH ME!

Learning to MULTIPLY Can be Fun!

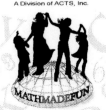

Math Made Fun
PO Box 4017
Ithaca, NY 14850
www.mathmadefun.com

This book is dedicated to my 2×2 daughters and my loving, supportive husband.

Acknowledgments

I am sincerely grateful to my editor Linda Glaser. Her numerous suggestions for restructuring the manuscript proved to be invaluable. I truly appreciate Laura Gates-Lupton for her many helpful ideas for improving the readability of the manuscript, and for providing copy editing. Thank you also to Mom, Dad, Bob, and Margaret for offering suggestions, proofreading and support. In addition, I appreciate the critique provided by the members of my critique group. Thanks to Laura Gates-Lupton, Jodie Mangor, Cathleen Banford, Johanna Husband, Andrea Hazard, and Sigrid Mortensen. Annie Zygarowicz also deserves gigantic thanks for doing an amazing job designing the cover and completing the interior graphics.

A six-year-old named Sarah deserves a special thanks for serving as my educational consultant. Her priceless grimaces, at key moments, showed me where to alter the text. Special thanks go to Sarah's three older sisters Lizzi, Emily, and Jessica, who at one time or another survived being my guinea pigs while I tested my ideas. I am so grateful to my four daughters and my husband Bruce for their love, patience, and understanding.

I appreciate the teachers and individuals who reviewed the book, wrote letters of recommendation or provided encouragement. Thanks to Ellen Mitterer, Renee Qamar, Lisa Neville, Rick Junge, Margaret Steinacher, Terri Stoff, Tracy Kirkman, Pam Wooster, Mary Psiaki, Jane Bruce-Robertson, Jacquie Lopez, Melody B., Pamela W., Nancy Reddy, Maria Rider, Nancy Saltzman, Bec Groves-Haley, Lisa Sanflippo, Karen Grace-Martin, Susan Hubbard, and Jessica Mitchell. Thank you also to the many other teachers and families who reviewed and tested the book.

Photo Credit: Jacqueline Conderacci © 2008

All rights reserved. No part of this book, may be reproduced, stored in a retrieval system, or transmitted in any form, on any means, electronic, mechanical, photocopying, recording or otherwise, without written consent of the publisher.

Math Made Fun
PO Box 4017
Ithaca, NY, 14850

www.mathmadefun.com

Printed in the United States of America

Introduction

When my daughter was learning to multiply, I encouraged her to jump on the trampoline, her favorite activity, while simultaneously practicing her "skip counting." She learned so quickly that we made jumping and skip counting a regular part of her "learning to multiply" routine. This resulted in her mastering multiplication in record time while enjoying the process immensely. I realized from this experience how important movement and rhythm are to learning math, and that was the spark that led to my writing this book.

Research has proven a correlation between movement and enhanced learning. At least thirteen different studies have confirmed that exercise stimulates brain growth. According to Carla Hannaford, PhD, author of *Smart Moves, Why Learning is Not All in Your Head*, "Movement integrates and anchors new information and experience into our neural networks…Moving while learning increases learning."

The ability to multiply quickly and correctly lays the foundation for future success in math. Confidence in math leads to self-assurance in other realms of a child's life. *Multiply With Me*: *Learning to Multiply Can Be Fun* provides an easy-to-use method for giving children exceptional multiplication skills. The activities included are fun, hands-on, physically active and child-centered. They are designed to meet the needs of multiple learning styles.

Math uses multiplication. When solving math problems, the math student must multiply, over and over again. Multiplication, also considered "fast addition," is the one math ability that makes or breaks a child's future achievement in math. He/she must be able to multiply before being asked to divide or to understand the relationships of fractions, decimals and percents. A child who can multiply will find it easy to learn how to work with least common multiples and greatest common factors. Multiplication is also crucial for developing competence in more advanced areas of math, including algebra.

Math is cumulative. If your child is insecure with multiplication, then learning new mathematical concepts might seem overwhelming. As the difficulty of math increases, multiplication is used more frequently. For example, in the following equation from a first level algebra course, multiplication could be used as many as five times before determining the answer: $5x^2 + 3x^3 = y$

Math requires confidence. A student with poor multiplication skills will experience great frustration when attempting to solve the problem above. A student may come to the conclusion that he/she is "bad at math." This lack of self-confidence leads to an unwillingness to try and often results in a self-perpetuating circle of failure in math. In comparison, a student with excellent multiplication skills solves this problem with ease and gains the feeling of self-esteem that comes from solving a complicated problem.

Math needs practice. Learning to multiply is comparable to learning to read. You must be able to read simple words before you can read entire passages. Practice is the key component in developing the ability to read and is just as important in math. In most schools, children are taught the mechanics of how to multiply and are given some practice time, but children need more practice to ensure a level of fluency in multiplying that will transfer into mathematical confidence. Asking a child to take algebra without this mathematical confidence is like asking a child to read *War and Peace* before he/she has mastered picture books.

Math begins at home. One of the most important contributions we can make to our children's competence in math is to teach them to multiply at the age of six or seven, before it is normally taught in school. Your child will then be practicing multiplication skills in school while others are learning them for the first time. This practice of skills already known will enhance your child's math confidence and facilitate his/her acquisition of more advanced math skills.

Math requires understanding. Too often in the past, the rote memorization of multiplication tables was required without explaining the numerical basis of multiplication. The initial use of flash cards to memorize multiplication tables stifles comprehension. Rote memorization creates a level of abstraction for which the child has not had adequate preparation, and thus lays the foundation for a child's frustration and misunderstanding of math. Instead of rote memorization, *Multiply With Me* uses a concrete method designed to enhance your child's understanding of multiplication. In addition, the method is enjoyable. After a child thoroughly understands the concepts of multiplication, flash cards may be helpful to increase speed.

Math requires perseverance. Be patient while your child figures out the relationships between numbers in multiplication. The time required for mastery varies between children just as it does in reading. Your child's understanding of math relationships will lead him/her to lifelong comfort with numbers and mathematics.

Multiplication needs to be a "tool" rather than a "task." Using multiplication as a tool means that a child is able to know the answer quickly and without anxiety. Consider how a child learns to walk. The beginning steps require a child's undivided attention, focus and concentration. Walking is a task. At some point, a child does not have to think about the mechanics of how to walk; he/she is able to walk and engage in other activities simultaneously. At this point, walking is a tool. If multiplication does not become a tool for your child before the sixth grade, then your child is at risk to develop math phobia, math anxiety, or to come to the conclusion that he/she is "bad at math." Multiplication becomes a tool when a child can complete one hundred single-digit multiplication problems in less than four minutes.

Multiplication with Groups or in the Elementary School Classroom
The techniques used in *Multiply With Me* are equally successful in groups or in the elementary school classroom. I recommend using non-food items (such as stickers) for counters when working with groups. Be sure to have your students whisper or shout the numbers at exactly the same time when counting together.

Multiply With Me is a complete, scripted course. You can simply read what you are instructed to say and how your child should answer. Some children will require the entire program as presented in *Multiply With Me*, while others will need only a fraction of the lessons. Modify the script and the lessons to meet your child's individual needs.

Multiply With Me covers the multiplication tables up to twelve times twelve. Your child should know how to count to 144 and be able to write the numbers up to 144 before starting this program. Perfect knowledge of addition and subtraction are not necessary. These skills will grow exponentially as your child completes the program.

The book is divided into forty-five lessons, each taking twenty to thirty minutes. Each lesson builds on the previous lessons. About half of the lesson time involves movement and hands-on activities while the other half is devoted to practicing multiplication in the workbook. Common household items such as pennies or food can be used for counting so that preparation time is minimized. The following chart summarizes each lesson. As you can see, the lessons are organized in sets of four, with lesson 33 as the exception.

Lessons	Covers Multiplication With	And Review of
1-4	2's	
5-8	3's	2's
9-12	4's	2's and 3's
13-16	5's	2's - 4's
17-20	6's	2's - 5's
21-24	7's	2's - 6's
25-28	8's	2's - 7's
29-32	9's	2's - 8's
33	Place Value	2's - 9's
34-37	10's	2's - 9's
38-41	11's	2's - 10's
42-45	12's	2's - 11's

One approach is to complete four lessons a week over an eleven-week period. If you choose this approach, I recommend reviewing the skip counting found at the end of the book with your child on days that you are not doing lessons. In addition, it is helpful to practice skip counting spontaneously, such as while in the car, before bed, first thing in the morning, when skipping rope, or doing any other form of physical activity. Note that Lesson 33 may be completed at any time. It may be useful to go through this lesson more than once.

Jump in and enjoy this delightful counting experience with your child!

Suzy Koontz

Ithaca, New York
March 21, 2008

How to Prepare For the Lessons

Determine a Name for Your Lesson Time

The time you spend teaching your child to multiply needs to have its own appropriate name. Just as we remind our children that it is time to practice the piano, clean their room, or do homework, I recommend calling these lessons *"time to practice multiplication"* or *"time to do our counting work."* The script uses the phrase *"time to do our counting work"* since multiplication is the tool that allows us to count very fast. The lessons in *Multiply With Me* prepare a child to multiply, by practicing counting.

Decide on Sorting/Counting Container or Compartment

Small trays are fine for lessons five through twenty where your child learns to multiply by threes, fours, and fives. Egg cartons are good for sorting and counting non-food items. Food items may be sorted on a clean tablecloth, in bowls, in sterilized plastic refrigerator egg containers, or in an ice cube tray. In order to have a twelve-compartment container, put masking tape over the extra compartments of an ice cube tray.

As the number of counters increases, your child will need more room to lay them out. For lessons five through forty-five, it is possible to lay out the counters on the table in front of you in twelve groups or piles, but I suggest using a clean kitchen tray that has been modified with masking tape. Use the masking tape to make three rows and four columns on the tray. Place the tape on the tray so that twelve equal compartments are created. The following chart shows how the tray should look:

Decide on Counting Items

Ordinary household items such as pennies, paper clips, small blocks or toothpicks may be used for counting. Food items such as gummy bears, chocolate covered raisins, small candies, sweet cereal, or pretzels generally provide an added incentive to your child, especially if he/she is allowed to eat some of them after the lesson.

The instructions in this program use gummy bears for counting. If you choose to use a different counter, simply substitute the name of that item in the script.

I recommend changing the counters often. New kinds of counters keep your child looking forward to lesson time. Children find the touch and the smell of the food items very appealing. The promise that he/she may eat some treats afterward helps a child to stay focused during lesson time.

In each lesson, the last two lines of the script prepare your child for eating some counters. These lines generally are, *"There is just one more thing to do before we eat a few of our counters,"* and *"Wonderful work! Now let's have a treat."* If you choose to use non-food counters, these lines should be deleted.

Decide on Time and Place to Do the Lesson

Each lesson is designed to take twenty to thirty minutes. Decide in advance when and where you will do the lessons. Try to make it the same time and place every day. One time that works well for families is immediately following the evening meal. This way the counting treats may be eaten for dessert.

Create a Calm, Quiet Atmosphere

Try to work in an uncluttered space. Turn off the television, stereo, computer and telephone. Try to minimize interruptions of any kind during your lesson time. Other children should be occupied with something or supervised by someone else so they don't interrupt. However, at the end of each lesson, encourage the entire family to join the skip-counting fun.

Focus on Your Child

Try to forget that you have bills to pay, dishes to wash, laundry and myriad other chores waiting. Focus all of your attention on your child and the lesson.

Communicate a Positive Message

Let your body language say "I am happy to have some special time with you, I love you, I love to spend time with you, and it is fun to count and practice multiplication with you." Smile, make eye contact, sit near your child. Be silly and laugh together.

Remember To Praise

Think in advance of ways to praise your child while working through the lesson. Add your own positive comments to the scripted dialogue.

Leave Behind Your Own Baggage

Maybe you have had bad experiences and don't think you are good at math. Maybe you hate math, or have a math phobia or math anxiety. Your child doesn't need to know any of this. It is easy for a child to adopt a parent's dislike or fear concerning math. Your child will benefit from your positive attitude about the advantages of practicing math.

Read In Advance

Quickly read through the complete program so that you have a general sense of the purpose and organization of each lesson and how the lessons build on each other.

Afterwards, spend time reading the first lesson carefully before you work with your child. The first lesson lays the foundation for the entire program and contains basic instructions that continue throughout the program.

General Information

The words in *italics* are what you say to your child. The plain-face words in smaller print describe the activity and how your child should answer.

Modify the Program as Necessary

Every child is different. You know your child best. Modify the script and the lesson to meet your child's specific needs. Following are some examples of how parents have modified the script to meet their child's needs or interests.

- A mother chose to use stickers as counters. She created a 12 by 12 grid on a sheet of paper to place the stickers and write the skip counting numbers.
- A mother chose to discontinue playing catch after Lesson 9.
- A mother chose to add whisper/loud to the stepping/clapping at the beginning of each lesson. One-to-one correspondence was solid for her child.
- A father chose to change "Mommy Bear" to "Daddy Bear" in the Sleeping Bears game.
- A mother discovered that her child only needed to complete the first twelve lessons, practice skip counting and do the workbook to learn to multiply.
- A mother found that her child enjoyed the program more when he was allowed to choose the counters.
- A classroom teacher used the program with her class by substituting "teacher" for the word "parent" in the script.

Every child has unique needs. Use all or any part of this program to assist you in helping your child learn to multiply. Your child will benefit from completing any of the lessons.

Items Needed Before Beginning the Program

- Counters such as gummy bears, crackers, stickers, pennies, or buttons
- Counting container
- Paper
- Pencil, pen and washable magic marker
- Masking tape
- Something to catch, such as a ball, beanie baby or balloon.
- Twelve dimes and nine pennies (not needed until Lesson 33)

Lesson 1

In This Lesson You Will:

1. Count with your child to 24 while stepping/clapping
2. Count with your child to 24 while jumping/clapping
3. Count with your child to 24 while playing catch
4. Have your child place 2 objects in each of the 12 compartments
5. Model "counting fingers"
6. Model counting objects by 1 with "counting fingers"
7. Count objects together by 1 with "counting fingers"
8. Have your child count objects by 1 with "counting fingers"
9. Model skip counting by 2's
10. Teach the "whisper/loud" counting trick
11. Count objects together using the "whisper/loud" counting trick
12. Have your child count objects using the "whisper/loud" counting trick
13. Have your child complete the workbook lesson
14. Recite or sing the skip counting chant together with any siblings or friends

Parent: *Now it is time to do our counting work. We will start with counting our steps. We will count to 24 while stepping and clapping. Let me show you: Take one step, clap and say one all at the same time. Take another step, clap and say two. Take another step, clap and say three. Let's do it together. Ready? Let's count! 1, 2, 3...24.*

Repeat the process until your child can do all 24 steps using one-to-one correspondence. It is essential that children develop skills in one-to-one correspondence. One-to-one correspondence is the ability to link a number name with one and only one object. The activity of counting steps is the first activity in all 45 lessons since it reinforces one-to-one correspondence. It also serves as a warm-up activity, preparing your child to concentrate and encouraging your child to stay focused.

Parent: *Now we will count with jumping instead of stepping. We will count to 24 while jumping and clapping. Jump (with both feet), clap and say 1 all at the same time. Jump, clap and say 2. Jump, clap and say 3. Ready? Jump! 1, 2, 3...24*

Parent: *Now we will count with catching instead of jumping. We will count to 24 while we toss and catch this beanbag. We will* *count as we catch the beanbag. Get ready. Here it comes: 1, 2, 3...24*

Continue to 24 tosses/attempted catches. Beanbags work well for this but feel free to toss a beanie baby, a ball, a balloon, or whatever is handy. Don't worry about your child not catching the item. Continue to count to 24 whether the item is caught or not.

Parent: *Now it is time to count the gummy bears. Take 24 gummy bears and put 2 gummy bears into each compartment.*

Child should put 2 gummy bears into each compartment.

Parent: *Now it is my turn to count by 1's. I need my counting fingers. Have you seen my counting fingers? Where are they? Oh, here they are!*

The counting fingers are the index and middle finger pressed together. Pull out your index and middle finger of your preferred hand from behind your back. Be as dramatic as you choose. The sillier you are the better. Using the index and middle finger pressed together rather than one finger for counting speeds up the learning process. Counting is the primary activity of this program. In all subsequent lessons, use the counting fingers every time counting is required.

Parent: *Where are your counting fingers?*

Help your child find his/her counting fingers. If your child is right-handed, then encourage him/her to count with the index and middle finger of his/her right hand. If he/she is left-handed, then encourage him/her to count with the index and middle fingers from his/her left hand.

Parent: *Now it is my turn to count by 1's. Watch me count by 1's using my counting fingers: 1, 2, 3...24.*

Be sure to use your counting fingers and point to each gummy bear as you count from 1 to 24. This models for your child the concept of one-to-one correspondence.

Parent: *Now we'll count by 1's. Are your counting fingers ready? From now on, when I say "get ready" it means to get your counting fingers ready. So get ready! Count: 1, 2, 3...24.*

Have your child count along with you. Be sure that he/she uses his/her index and middle finger and points to each gummy bear as he/she says the numbers.

Parent: *Now it is your turn to count by 1's. Get ready. Count.*

Child counts from 1 to 24 by him/herself, pointing with counting fingers to each gummy bear as he/she counts.

Parent: *Now it is my turn to count by 2's. I will count two at a time. Did you know that any two items is called a pair? Watch me count.*

Use your counting fingers. Be sure to point to each pair as you count.

Parent: *2, 4, 6, 8, 10, 12, 14, 16, 18, 20, 22, 24. I bet you are wondering how I got to be such a good counter. I want to share my counting secret with you. Let me whisper it in your ear. Shh.*

Put your counting fingers to your lips and say *shhh* to emphasize that this is a secret.

Parent: *I used the whisper/loud counting trick. I will count the gummy bears more slowly so that I can show you the secret counting trick. I will point to each gummy bear as I count it. When I point to the first gummy bear, I will whisper 1 so quietly that you will hardly be able to hear me. When I point to the second gummy bear, I will say 2 out LOUD. When I point to the third gummy bear, I will whisper 3. When I point to the fourth gummy bear, I will say 4 out LOUD. Watch me count!*

1 (whisper it)
2 (say it LOUD)
3 (whisper it)
4 (say it LOUD)
5 (whisper it)
6 (say it LOUD)
7 (whisper it)

8 (say it LOUD)
9 (whisper it)
10 (say it LOUD)
11 (whisper it)
12 (say it LOUD)
13 (whisper it)
14 (say it LOUD)
15 (whisper it)
16 (say it LOUD)
17 (whisper it)
18 (say it LOUD)
19 (whisper it)
20 (say it LOUD)
21 (whisper it)
22 (say it LOUD)
23 (whisper it)
24 (say it LOUD)

Parent: *Now we'll count these 24 gummy bears together. Let's use our counting fingers and our secret counting trick. (Shh). Get ready. Count. 1 (whisper), 2 (LOUD), 3 (whisper), 4 (LOUD), 5 (whisper), 6 (LOUD), 7 (whisper), 8 (LOUD), 9 (whisper), 10 (LOUD), 11 (whisper), 12 (LOUD), 13 (whisper), 14 (LOUD), 15 (whisper), 16 (LOUD), 17 (whisper), 18 (LOUD), 19 (whisper), 20 (LOUD), 21 (whisper), 22 (LOUD), 23 (whisper), 24 (LOUD). Terrific counting!*

Parent: *Now it is your turn to count using our secret counting trick. Get ready! Count!*

Child should count to 24 using the whisper/loud technique and his/her counting fingers. It is common for children to mix up the volume in this activity. For example, he/she may whisper the 2 when he/she intends to say it loud or he/she may say the 3 loud when he/she intends to whisper it. Allow your child to self-correct his/her error rather than correcting him/her. Reassure your child and laugh with him/her over the error. Model the activity again if your child doesn't catch his/her own mistake.

The main purpose of using the whisper/loud counting method is to show your child that when we skip count (i.e. count 2, 4, 6, 8, 10...), we still are counting each item. Repeat this activity if necessary. The whisper/loud counting is used in nearly all the lessons. It is very helpful in understanding skip counting.

Parent: *Now it is time to do some pages in the workbook.*

Child should complete Lesson 1 in the workbook.

Parent: *Fantastic counting! There is just one more thing to do before we eat a few of our counters. We will recite the skip counting chant!*

Recite or sing together the skip counting chant found at the end of this book. (Have your child color in the skip counting patterns of numbers according to the directions.) If you choose to sing the skip counting chant, then make up your own tunes according to songs already familiar to your child. It is best to get up and move about while reciting. Use any type of physical movement that is appropriate for your environment such as dancing; running or jumping in place; jumping on an outdoor trampoline; turning in circles; doing jumping jacks; riding on a pretend horse; doing cartwheels, somersaults or handstands; or anything that your comfort and space permit.

Try to be as silly as possible while reciting or singing. Use different voices; speak in soft tones or shout as loud as possible. The purpose of learning the skip counting chant is to make your child familiar with the pattern of the skip counting numbers. Each week your child will learn to multiply by a new number. Your child will learn to multiply more quickly if he/she is familiar with these patterns of numbers. This is especially true in later lessons. For this reason, the activity of reciting or singing the skip counting chant is at the end of each of the 45 lessons. I recommend altering the type of physical movement daily or weekly to keep your child interested.

If you have other children, then it is beneficial for them to participate in this part of the lesson. Even if your older children are learning more advanced math concepts, the skip counting chant will still be helpful. Learning the skip counting chant will help your older child learn to how to divide, find least common denominators and greatest common factors and facilitate the process of learning algebra. Younger children listening and participating will have a huge advantage when it is their turn to learn to multiply.

Lesson 2

In This Lesson You Will:

1. Count with your child to 24 while stepping/clapping
2. Count with your child to 24 while jumping/clapping
3. Count with your child to 24 while catching
4. Have your child place 2 objects in each of the 12 compartments
5. Model counting objects by 1 using the "whisper/loud" counting trick
6. Count objects together by 1 using the "whisper/loud" counting trick
7. Have your child counts objects by 1 using the "whisper/loud" counting trick
8. Model skip counting by 2's
9. Skip count together by 2's
10. Play the sleeping bears game
11. Have your child skip count by 2's
12. Play the Add On 2 game
13. Have your child complete the workbook lesson
14. Recite or sing the skip counting chant together with any siblings or friends

Parent: *Let's start with counting our steps. We will count to 24 while stepping and clapping. Take one step, clap and say one all at the same time. Take another step, clap and say two. Take another step, clap and say three. Let's start at 1. Ready? Count! 1, 2, 3...24*

Parent: *Now we will count with jumping instead of stepping. We will count to 24 while jumping and clapping. Jump (with both feet), clap and say 1. Jump, clap and say 2. Jump, clap and say 3. Let's start at 1. Ready? Jump! 1, 2, 3...24*

Parent: *Now we will count with catching instead of jumping. We will count to 24 while we toss and catch this beanbag. We will count as we catch the beanbag. Get ready. Here it comes: 1, 2, 3...24*

Parent: *Now it is time to count the gummy bears. Put 2 gummy bears into each compartment.*

Parent: *Now it is my turn to count the gummy bears using our secret counting method. Shh. I need my counting fingers. Have you seen my counting fingers? Where are they? Oh, here they are. Now I will count. 1 (whisper), 2 (LOUD), 3 (whisper), 4 (LOUD), 5 (whisper), 6 (LOUD), 7 (whisper), 8 (LOUD), 9 (whisper), 10 (LOUD), 11 (whisper), 12 (LOUD), 13 (whisper), 14 (LOUD), 15 (whisper), 16 (LOUD), 17 (whisper), 18 (LOUD), 19 (whisper), 20 (LOUD), 21 (whisper), 22 (LOUD), 23 (whisper), 24 (LOUD).*

Parent: *Now we'll count these 24 gummy bears together, using our secret counting trick. (Shh). Are your counting fingers ready? Get ready. Count. 1, 2, 3, 4, 5, 6, 7, 8, 9, 10, 11, 12, 13, 14, 15, 16, 17, 18, 19, 20, 21, 22, 23, 24.*

Help your child to find his/her counting fingers. Be sure that he/she uses his/her index and middle finger and points to each compartment as he/she says the numbers. Use the whisper/loud counting technique.

Parent: *Now it is your turn to count these 24 gummy bears. Use our secret counting trick and your counting fingers. Get ready. Count.*

Child counts to 24 using the whisper/loud technique and his/her counting fingers.

Parent: *Now I will skip count the gummy bears by 2's. This is called skip counting by 2's because I will skip all the numbers that I whispered when we counted with our secret counting trick. Skip counting sounds the same as our secret counting trick but it can be much, much faster. Watch how fast I can skip count these gummy bears. 2, 4, 6, 8, 10, 12, 14, 16, 18, 20, 22, 24.*

Be sure to point to the group of 2 when you model skip counting. This helps your child to understand that when he/she skip counts, he/she is adding on 2 at a time.

Parent: *Now it is our turn to skip count by 2's. Are your counting fingers ready? Get ready. Count: 2, 4, 6, 8, 10, 12, 14, 16, 18, 20,*

22, 24.

Have your child count along with you. Be sure that he/she uses his/her index and middle finger and points to each gummy bear as he/she says the numbers.

If your child can count by 2's easily, then skip the Sleeping Bears Game.

THE SLEEPING BEARS GAME

Dump out the counting container.

Parent: *Now we will play a game called Sleeping Bears. We each take turns being the Mommy Bear counting our sleeping bear cubs. When we put the bear cubs in the compartment, we will pretend that we are tucking them in bed. We will tuck 2 bears in bed at a time. They fall asleep immediately! Then the Mommy needs to know how many of her little bear cubs are asleep so she counts them all up. You can be the Mommy Bear first.*

Take these gummy bears; put them in the first compartment. How many are there?

Have your child take 2 gummy bears, put them in the first compartment and say, "2 sleeping bear cubs."
Parent puts 2 gummy bears in the next available compartment and says, "2, 4 sleeping bear cubs."

Parent: *Now your turn.*

Child puts 2 gummy bears in the next available compartment and says, "2, 4, 6 sleeping bear cubs."
Parent puts 2 gummy bears in the next available compartment and says, "2, 4, 6, 8 sleeping bear cubs."
Continue until all compartments are filled. If your child likes the game enough, play it again.

Parent: *Now it is your turn to count by 2's. Get ready. Count.*

Child skip counts by 2's from 2 to 24, pointing with counting fingers to each compartment of 2 gummy bears.

If your child has difficulty counting by twos, count together and review this lesson.

Parent: *Excellent counting!*

THE ADD ON 2 GAME

Parent: *Now will play the Add On 2 game. Did you know that when we skip count by 2's, we start at zero and add on 2? Then we keep adding on 2 to the previous number. Watch me add up all the gummy bears by adding on 2.*

Begin by pointing to an empty compartment to explain 0 or no gummy bears. Then point to each compartment as you add on the 2 gummy bears.

Parent: *0 or no gummy bears plus 2 gummy bears is equal to 2 gummy bears.*
2 gummy bears plus 2 gummy bears is equal to 4 gummy bears.
4 gummy bears plus 2 gummy bears is equal to 6 gummy bears.
6 gummy bears plus 2 gummy bears is equal to 8 gummy bears
8 gummy bears plus 2 gummy bears is equal to 10 gummy bears
10 gummy bears plus 2 gummy bears is equal to 12 gummy bears
12 gummy bears plus 2 gummy bears is equal to 14 gummy bears
14 gummy bears plus 2 gummy bears is equal to 16 gummy bears
16 gummy bears plus 2 gummy bears is equal to 18 gummy bears
18 gummy bears plus 2 gummy bears is equal to 20 gummy bears
20 gummy bears plus 2 gummy bears is equal to 22 gummy bears
22 gummy bears plus 2 gummy bears is equal to 24 gummy bears

Parent: *Now we'll skip count by 2's together by adding on 2.*

Repeat previous step.

Parent: *Now it is your turn to skip count by 2's by adding on 2.*

Repeat step above.

Parent: *Now it is time to do some pages in the workbook.*

Child completes Lesson 2 in the workbook.

Parent: *There is just one more thing to do before we eat a few of our counters! We will recite the skip counting chant!*

Recite the skip counting chant using some type of physical movement.

Parent: *Wonderful work. Now let's have a treat!*

Lesson 3

In This Lesson You Will:

1. Count with your child to 24 while stepping/clapping
2. Count with your child to 24 while jumping/clapping
3. Count with your child to 24 while catching
4. Have your child place 2 objects in each of the 12 compartments
5. Model counting objects by 1 using the "whisper/loud" counting trick
6. Count objects together by 1 using the "whisper/loud" counting trick
7. Have your child count objects by 1 using the "whisper/loud" counting trick
8. Model skip counting by 2's
9. Skip count together by 2's
10. Have your child skip count by 2's
11. Play the Add On 2 game
12. Play the How Many? game
13. Have your child complete the workbook lesson
14. Recite or sing the skip counting chant with any siblings or friends

Parent: *Let's start with counting our steps. We will count to 24 while stepping and clapping. Ready? Count! 1, 2, 3...24.*

Parent: *Now we will count with jumping instead of stepping. We will count to 24 while jumping and clapping. Ready? Jump: 1, 2, 3...24.*

Parent: *Now we will count with catching instead of jumping. We will count to 24 while we toss and catch this beanbag. We will count as we catch the beanbag. Get ready. Here it comes: 1, 2, 3...24.*

Parent: *Now it is time to count the gummy bears. Put 2 gummy bears into each compartment.*

Parent: *Now it is my turn to count the gummy bears using our* *secret counting method. Shh. I need my counting fingers. Have you seen my counting fingers? Where are they? Oh, here they are. Now I will count. 1 (whisper), 2 (LOUD), 3 (whisper), 4 (LOUD), 5 (whisper), 6 (LOUD), 7 (whisper), 8 (LOUD), 9 (whisper), 10 (LOUD), 11 (whisper), 12 (LOUD), 13 (whisper), 14 (LOUD), 15 (whisper), 16 (LOUD), 17 (whisper), 18 (LOUD), 19 (whisper), 20 (LOUD), 21 (whisper), 22 (LOUD), 23 (whisper), 24 (LOUD).*

Parent: *Now we'll count these 24 gummy bears together, using our secret counting trick. (Shh). Are your counting fingers ready? Get ready. Count: 1, 2, 3, 4, 5, 6, 7, 8, 9, 10, 11, 12, 13, 14, 15, 16, 17, 18, 19, 20, 21, 22, 23, 24.*

Help your child to find his/her counting fingers. Be sure that he uses his/her index and middle finger and points to each compartment as he/she says the numbers. Use the whisper/loud counting technique.

Parent: *Now it is your turn to count these 24 gummy bears. Use our secret counting trick and your counting fingers. Get ready. Count.*

Child counts to 24 using the whisper/loud technique and his/her counting fingers.

Parent: *Now I will skip count the gummy bears by 2's. Remember that this is called skip counting by 2's because I will skip all the numbers that I whispered when we counted with our secret counting trick. Listen to how fast I can skip count these gummy bears! 2, 4, 6, 8, 10, 12, 14, 16, 18, 20, 22, 24.*

Parent: *Now we'll skip count by 2's together. Do you think that you can be fast like me? Are your counting fingers ready? Get ready. Count: 2, 4, 6, 8, 10, 12, 14, 16, 18, 20, 22, 24.*

Have your child count along with you. Be sure that he/she uses his/her counting fingers and points to each compartment as he/she says the numbers.

Parent: *Now it is your turn to count by 2's. Get ready. Count.*

Child skip counts by 2's from 2 to 24, pointing with counting fingers to each compartment of 2 gummy bears. Continue to practice this until your child can easily count by 2's.

Parent: *Excellent counting!*

THE ADD ON 2 GAME

Parent: *Do you remember that when we skip count by 2's, we start at zero and add on 2? Then we keep adding on 2 to the previous number. Watch me add up all the gummy bears by adding on 2.*

Begin by pointing to an empty compartment to explain 0 or no gummy bears. Then point to each compartment as you add on the 2 gummy bears.

Parent: *0 or no gummy bears plus 2 gummy bears is equal to 2 gummy bears.*
2 gummy bears plus 2 gummy bears is equal to 4 gummy bears.
4 gummy bears plus 2 gummy bears is equal to 6 gummy bears.
6 gummy bears plus 2 gummy bears is equal to 8 gummy bears
8 gummy bears plus 2 gummy bears is equal to 10 gummy bears
10 gummy bears plus 2 gummy bears is equal to 12 gummy bears
12 gummy bears plus 2 gummy bears is equal to 14 gummy bears
14 gummy bears plus 2 gummy bears is equal to 16 gummy bears
16 gummy bears plus 2 gummy bears is equal to 18 gummy bears
18 gummy bears plus 2 gummy bears is equal to 20 gummy bears
20 gummy bears plus 2 gummy bears is equal to 22 gummy bears
22 gummy bears plus 2 gummy bears is equal to 24 gummy bears

Parent: *Now we'll skip count by 2's together by adding on 2.*

Repeat previous step.

Parent: *Now it is your turn to skip count by 2's by adding on 2.*

Repeat previous step.

THE HOW MANY? GAME

Parent: *Now it is time to play a multiplication game called How Many? Put 2 gummy bears in the first compartment.*

Dump out the counting container and have your child fill one compartment with 2 gummy bears.

Parent: *How many compartments have gummy bears?*

Child should answer that 1 compartment is filled.

Parent: *How many gummy bears are there?*

Child should answer that there are 2 gummy bears.

Parent: *You just multiplied! 1 times 2 is 2. 1 compartment filled with 2 gummy bears is 2 gummy bears. Now fill up another compartment with 2 gummy bears.*

Child puts 2 gummy bears in another compartment.

Parent: *How many compartments have gummy bears?*

Child should answer that 2 compartments are filled.

Parent: *Use your skip counting to figure out how many gummy bears there are.*

Child should skip count and determine that there are 4 total gummy bears.

Parent: *You just multiplied again! 2 times 2 is 4. 2 compartments filled with 2 gummy bears are 4 gummy bears.*

Parent: *This is written as 2×2.*
Show your child on paper how 2 x 2 looks. Then show him/her that multiplication is also written as one number over the other with the multiplication or times symbol, ×, written

next to the bottom number.
For example,

$$\begin{array}{r} 2 \\ \times 2 \\ \hline \end{array}$$

Parent: *Now fill up another compartment with 2 gummy bears. How many compartments have gummy bears?*

Child should answer that 3 compartments are filled.

Parent: *How many gummy bears are there?*

Child should skip count and determine that there are 6 total gummy bears.

Parent: *You multiplied again! 3 times 2 is 6. 3 compartments filled with 2 gummy bears are 6 gummy bears. This is written as 3×2.*

Show your child on paper how 3 x 2 looks.

Parent: *Now fill up another compartment with 2 gummy bears. How many compartments have gummy bears?*

Child should answer that 4 compartments are filled.

Parent: *How many gummy bears are there?*

Child should skip count and determine that there are 8 total gummy bears.

Parent: *Terrific multiplying! 4 times 2 is 8. 4 compartments filled with 2 gummy bears are 8 gummy bears. This is written as 4×2.*

Show your child on paper how 4 x 2 looks.

Parent: *Now fill up another compartment with 2 gummy bears. How many compartments have gummy bears?*

Child should answer that 5 compartments are filled.

Parent: *How many gummy bears are there?*

Child should skip count and determine that there are 10 total gummy bears.

Parent: *You multiplied again! 5 times 2 is 10. 5 compartments filled with 2 gummy bears are 10 gummy bears. This is written as 5×2.*

Show your child on paper how 5 x 2 looks.

Parent: *Now fill up another compartment with 2 gummy bears. How many compartments have gummy bears?*

Child should answer that 6 compartments are filled.

Parent: *How many gummy bears are there?*

Child should skip count and determine that there are 12 total gummy bears.

Parent: *Marvelous multiplying! 6 times 2 is 12. 6 compartments filled with 2 gummy bears are 12 gummy bears. This is written as 6×2.*

Show your child on paper how 6 x 2 looks.

Parent: *Now fill up another compartment with 2 gummy bears. How many compartments have gummy bears?*

Child should answer that 7 compartments are filled.

Parent: *How many gummy bears are there?*

Child should skip count and determine that there are 14 total gummy bears.

Parent: *You just multiplied again! 7 times 2 is 14. 7 compartments filled with 2 gummy bears are 14 gummy bears. This is written as 7×2.*

Show your child on paper how 7 x 2 looks.

Parent: *Now fill up another compartment with 2 gummy bears. How many compartments have gummy bears?*

Child should answer that 8 compartments are filled.

Parent: *How many gummy bears are there?*

Child should skip count and determine that there are 16 total gummy bears.

Parent: *I bet you didn't know that you could multiply so well! But you did it! 8 times 2 is 16. 8 compartments filled with 2 gummy bears are 16 gummy bears. This is written as 8×2.*

Show your child on paper how 8 x 2 looks.

Parent: *Now fill up another compartment with 2 gummy bears. How many compartments have gummy bears?*

Child should answer that 9 compartments are filled.

Parent: *How many gummy bears are there?*

Child should skip count and determine that there are 18 total gummy bears.

Parent: *Yes. Your multiplication is correct. 9 times 2 is 18. 9 compartments filled with 2 gummy bears are 18 gummy bears. This is written as 9×2.*

Show your child on paper how 9 x 2 looks.

Parent: *Now fill up another compartment with 2 gummy bears. How many compartments have gummy bears?*

Child should answer that 10 compartments are filled.

Parent: *How many gummy bears are there?*

Child should skip count and determine that there are 20 total gummy bears.

Parent: *You multiplied correctly again! 10 times 2 is 20. 10 compartments filled with 2 gummy bears are 20 gummy bears. This is written as 10×2.*

Show your child on paper how 10 x 2 looks.

Parent: *I am very impressed with your remarkable multiplying! Now it is time to do some pages in the workbook.*

Child completes Lesson 3 in the workbook.

Parent: *There is just one more thing to do before we eat a few of our counters! We will recite the skip counting chant!*

Recite the skip counting chant using some type of physical movement. Remember to move around and be silly while reciting. Include any siblings or friends.

Parent: *Wonderful work. Now let's have a treat!*

Lesson 4

In This Lesson You Will:

1. Count with your child to 24 while stepping/clapping
2. Count with your child to 24 while jumping/clapping
3. Count with your child to 24 while catching
4. Have your child place 2 objects in each of the 12 compartments
5. Model counting objects by 1 using the "whisper/loud" counting trick
6. Count objects together by 1 using the "whisper/loud" counting trick
7. Have your child count objects by 1 using the "whisper/loud" counting trick
8. Model skip counting by 2's
9. Skip count together by 2's
10. Have your child skip count by 2's
11. Play the How Many? game
12. Play the Finger Counting game
13. Model using fingers to solve multiplication problems
14. Practice solving multiplication problems together
15. Have your child solve multiplication problems
16. Have your child complete the workbook lesson
17. Recite or sing the skip counting chant together with any siblings or friends.

Parent: *Let's start with counting our steps. We will count to 24 while stepping and clapping. Ready? Count! 1, 2, 3...24.*

Parent: *Now we will count with jumping instead of stepping. We will count to 24 while jumping and clapping. Ready? Jump. 1, 2, 3...24.*

Parent: *Now we will count with catching instead of jumping. We will count to 24 while we toss and catch this beanbag. We will count as we catch the beanbag. Get ready. Here it comes. 1, 2, 3...24.*

Parent: *Now it is time to count the gummy bears. Put 2 gummy bears into each compartment.*

Parent: *Now it is my turn to count the gummy bears using our secret counting method. 1, 2, 3...24.*

Parent: *Now we'll count these 24 gummy bears together, using our secret counting trick. (Shh). Show me your counting fingers! Get ready. Count: 1, 2, 3...24.*

Parent and child count to 24 using the whisper/loud counting technique.

Parent: *Now it is your turn to count these 24 gummy bears with our secret counting trick and your counting fingers. Get ready. Count.*

Child counts to 24 using the whisper/loud counting technique and his/her counting fingers.

Parent: *Now I will skip count the gummy bears by 2's. Listen to how fast I can skip count these gummy bears! 2, 4, 6, 8, 10, 12, 14, 16, 18, 20, 22, 24.*

Parent: *Now we'll skip count by 2's together. Do you think that you can count fast like me? Are your counting fingers ready? Get ready. Count: 2, 4, 6, 8, 10, 12, 14, 16, 18, 20, 22, 24.*

Have your child count along with you. Be sure that he/she points to each compartment with his/her counting fingers as he/she skip counts.

Parent: *Now it is your turn to count by 2's. Get ready. Count.*

Child skip counts by 2's from 2 to 24 pointing with counting fingers to each compartment of 2 gummy bears. Continue to practice this until your child can easily count by 2's.

Parent: *Excellent counting!*

THE HOW MANY? GAME

Parent: *Now we will play our How Many? multiplication game. Put 2 gummy bears in the first compartment.*

Dump out the counting container and have your child fill one compartment with 2 gummy bears.

Parent: *How many compartments have gummy bears?*

Child should answer that 1 compartment is filled.

Parent: *How many gummy bears are there?*

Child should answer that there are 2 gummy bears.

Parent: *You just multiplied! 1 times 2 is 2. 1 compartment filled with 2 gummy bears is 2 gummy bears. Now fill up another compartment with 2 gummy bears.*

Child puts 2 gummy bears in another compartment.

Parent: *How many compartments have gummy bears?*

Child should answer that 2 compartments are filled.

Parent: *How many gummy bears are there?*

Child should skip count and determine that there are 4 total gummy bears.

Parent: *You just multiplied again! 2 times 2 is 4. 2 compartments filled with 2 gummy bears are 4 gummy bears.*

Parent: *This is written as 2×2.*

Show your child on paper how 2 x 2 looks. Then show him/her that multiplication is also written as one number over the other with the multiplication or times symbol, ×, written next to the bottom number. For example,

$$\begin{array}{r} 2 \\ \times 2 \\ \hline \end{array}$$

Parent: *Now fill up another compartment with 2 gummy bears. How many compartments have gummy bears?*

Child should answer that 3 compartments are filled.

Parent: *How many gummy bears are there?*

Child should skip count and determine that there are 6 total gummy bears.

Parent: *You multiplied again! 3 times 2 is 6. 3 compartments filled with 2 gummy bears are 6 gummy bears. This is written as 3×2.*

Show your child on paper how 3 x 2 looks.

Parent: *Now fill up another compartment with 2 gummy bears. How many compartments have gummy bears?*

Child should answer that 4 compartments are filled.

Parent: *How many gummy bears are there?*

Child should skip count and determine that there are 8 total gummy bears.

Parent: *Terrific multiplying! 4 times 2 is 8. 4 compartments filled with 2 gummy bears are 8 gummy bears. This is written as 4×2.*

Show your child on paper how 4 x 2 looks.

Parent: *Now fill up another compartment with 2 gummy bears. How many compartments have gummy bears?*

Child should answer that 5 compartments are filled.

Parent: *How many gummy bears are there?*

Child should skip count and determine that there are 10 total gummy bears.

Parent: *You multiplied again! 5 times 2 is 10. 5 compartments filled with 2 gummy bears are 10 gummy bears. This is written as 5×2.*

Show your child on paper how 5 x 2 looks.

Parent: *Now fill up another compartment with 2 gummy bears. How many compartments have gummy bears?*

Child should answer that 6 compartments are filled.

Parent: *How many gummy bears are there?*

Child should skip count and determine that there are 12 total gummy bears.

Parent: *Spectacular multiplying! 6 times 2 is 12. 6 compartments filled with 2 gummy bears are 12 gummy bears. This is written as 6×2.*

Show your child on paper how 6 x 2 looks.

Parent: *Now fill up another compartment with 2 gummy bears. How many compartments have gummy bears?*

Child should answer that 7 compartments are filled.

Parent: *How many gummy bears are there?*

Child should skip count and determine that there are 14 total gummy bears.

Parent: *You just multiplied again! 7 times 2 is 14. 7 compartments filled with 2 gummy bears are 14 gummy bears. This is written as 7×2.*

Show your child on paper how 7 x 2 looks.

Parent: *Now fill up another compartment with 2 gummy bears. How many compartments have gummy bears?*

Child should answer that 8 compartments are filled.

Parent: *How many gummy bears are there?*

Child should skip count and determine that there are 16 total gummy bears.

Parent: *I bet you didn't know that you could multiply so well! But you did it! 8 times 2 is 16. 8 compartments filled with 2 gummy bears are 16 gummy bears. This is written as 8×2.*

Show your child on paper how 8 x 2 looks.

Parent: *Now fill up another compartment with 2 gummy bears. How many compartments have gummy bears?*

Child should answer that 9 compartments are filled.

Parent: *How many gummy bears are there?*

Child should skip count and determine that there are 18 total gummy bears.

Parent: *Yes. Your multiplication is correct. 9 times 2 is 18. 9 compartments filled with 2 gummy bears are 18 gummy bears. This is written as 9×2.*

Show your child on paper how 9 x 2 looks.

Parent: *Now fill up another compartment with 2 gummy bears. How many compartments have gummy bears?*

Child should answer that 10 compartments are filled.

Parent: *How many gummy bears are there?*

Child should skip count and determine that there are 20 total gummy bears.

Parent: *You multiplied correctly again! 10 times 2 is 20. 10 compartments filled with 2 gummy bears are 20 gummy bears. This is written as 10×2.*

Show your child on paper how 10 x 2 looks.

Parent: *I am very impressed with your amazing multiplying!*

THE FINGER COUNTING GAME

Parent: *Now it is time to play a game called Finger Counting. Let's put 2 dots on each of my fingers.*

Put 2 dots on the backs of your fingers, on or near the fingernails, with a washable magic marker. If you don't like magic marker on your hands, then you can make dots with small round stickers or use a hand-held hole puncher to punch holes in masking tape. Hold out your fingers and wiggle them.

Parent: *When I skip count, I will be counting the dots on my fingers. Watch me skip count or count the dots on my fingers.*

Skip count on your fingers by gently squeezing the top of your finger with the thumb and index finger of your other hand. Start with the pinky on your left hand.
Squeeze the top of your pinky and say 2.
Squeeze the top of your ring finger and say 4.
Squeeze the top of your middle finger and say 6.
Squeeze the top of your index finger and say 8.
Squeeze the top of your thumb and say 10.
Switch hands and squeeze the top of your right thumb while saying 12. Count 14, 16, 18 and 20 on your right index, middle, ring finger and pinky, consecutively.

Parent: *Now I need to borrow 2 of your fingers. I need to count dots on your fingers as well. Let's put 2 dots on 2 of your fingers.*
Put 2 dots on each of your child's fingers with a washable magic marker, or with masking tape dots as described above.

Parent: *Now I will count the dots on my fingers and your fingers. Watch me skip count on our fingers.*

Skip count on your fingers again as described above. Borrow 2 fingers from your child to count 22 and 24.

Parent: *Now it is your turn to play Finger Counting. Wiggle your fingers. Skip count on your fingers.*

Help your child to skip count by 2's on his/her fingers. Use 2 of your fingers for 22 and

24. Practice 3 times or until your child is comfortable with Finger Counting.

Parent: *I have a secret for you! Fingers are great for solving multiplication problems! I will show you how to use fingers to multiply.*

Parent: *To find out the answer to 3×2, I'll hold up 3 fingers and skip count 3 times by 2's: 2, 4, 6.*

Hold up 3 fingers and really emphasize squeezing the top of each finger while counting 2, 4, 6.

Parent: *This is the same as counting the 2 dots on each of my 3 fingers. The answer to 3×2 is 6.*

Parent: *We may not always put dots on our fingers to practice skip counting. If we don't have dots on our fingers, then we will pretend that we do when we practice skip counting on our fingers.*

Parent: *Now we'll figure out 4×2 together. Hold up 4 fingers and skip count by 2's: 2, 4, 6, 8. The answer to 4×2 is 8.*

Help your child to hold up 4 fingers and use his/her fingers to skip count: 2, 4, 6, 8.

Parent: *Now it's your turn to figure out 3×2. Hold up 3 fingers and skip count.*

Help your child to hold up 3 fingers and skip count. Child should answer that 3 times 2 is 6.

Parent: *Now it's your turn to figure out 4×2. Hold up 4 fingers and skip count.*

Help your child to hold up 4 fingers and skip count. Child should answer that 4 times 2 is 8.

Parent: *What about 5×2? Hold up 5 fingers and skip count.*

Child holds up 5 fingers and skip counts. Child should answer that 5 times 2 is 10.

Parent: *What about 6×2? Hold up 6 fingers and skip count.*

Child holds up 6 fingers and skip counts. Child should answer that 6 times 2 is 12.

Parent: *What about 7×2? Hold up 7 fingers and skip count.*

Child holds up 7 fingers and skip counts. Child should answer that 7 times 2 is 14.

Parent: *What about 8×2? Hold up 8 fingers and skip count.*

Child holds up 8 fingers and skip counts. Child should answer that 8 times 2 is 16.

Parent: *What about 9×2? Hold up 9 fingers and skip count.*

Child holds up 9 fingers and skip counts. Child should answer that 9 times 2 is 18.

Parent: *What about 10×2? Hold up 10 fingers and skip count.*

Child holds up 10 fingers and skip counts. Child should answer that 10 times 2 is 20.

Parent: *Now you know how to multiply by 2's! Let's do some pages in the workbook.*

Child completes Lesson 4 in the workbook.

Parent: *There is just one more thing to do before we eat a few of our counters! We will recite the skip counting chant!*

Recite the skip counting chant using some type of physical movement. Remember to move around and be silly while reciting. Include any siblings or friends.

Parent: *Wonderful work. Now let's have a treat!*

Lesson 5

In This Lesson You Will:

1. Count with your child to 36 while stepping/clapping
2. Count with your child to 36 while jumping/clapping
3. Count with your child to 36 while playing catch
4. Have your child place 3 objects in each of the 12 compartments
5. Model counting objects by 1 with "counting fingers"
6. Count objects together by 1 with "counting fingers"
7. Have your child count objects by 1 with "counting fingers"
8. Model skip counting by 3's
9. Model the "whisper/loud" counting trick
10. Count objects together using the "whisper/loud" counting trick
11. Have your child count objects using the "whisper/loud" counting trick
12. Have your child complete the workbook lesson
13. Recite or sing the skip counting chant together with any siblings or friends.

Parent: *Let's start with counting our steps. We will count to 36 while stepping and clapping. Ready? Count! 1, 2, 3...36.*

Parent: *Now we will count with jumping instead of stepping. We will count to 36 while jumping and clapping. Ready? Jump: 1, 2, 3...36.*

Parent: *Now we will count with catching instead of jumping. We will count to 36 while we toss and catch this beanbag. We will count as we catch the beanbag. Get ready. Here it come: 1, 2, 3...36.*

Parent: *Now it is time to count the gummy bears. Take 36 gummy bears and put 3 gummy bears into each of the 12 compartments.*

Child should put 3 gummy bears into each compartment.

Parent: *Now it is my turn to count by 1's. I need my counting fingers. Have you seen my counting fingers? Where are they?*

Oh, here they are!

Pull out your counting fingers from behind your back.

Parent: *Watch me count by 1's using my counting fingers. 1, 2, 3…36.*

Be sure to use your counting fingers and point to each gummy bear as you count from 1 to 36. This models the concept of one-to-one correspondence.

Parent: *Now we'll count by 1's together. Are your counting fingers ready? Get ready. Count: 1, 2, 3…36.*

Have your child count along with you. Be sure that he/she uses his/her counting fingers and points to each gummy bear as he/she says the numbers. Repeat the activity if your child is still having difficulty with one-to-one correspondence.

Parent: *Now it is your turn to count by 1's. Get ready. Count.*

Child counts from 1 to 36, pointing to each gummy bear as he/she counts.

Parent: *Excellent counting. Now it is my turn to count by 3's. Watch me count.*

Use your counting fingers. Point to each compartment as you count.

Parent: *3, 6, 9, 12, 15, 18, 21, 24, 27, 30, 33, 36. Do you remember why I am such a good counter?*

Parent: *I use the whisper/loud counting trick. I will count the gummy bears more slowly so that I can show you the secret counting trick. I will point to each gummy bear as I count it. When I point to the first gummy bear, I will whisper 1 so quietly that you will hardly be able to hear me. When I point to the second gummy bear, I will whisper 2 so quietly that you will hardly be able to hear me. When I point to the third gummy bear, I will say 3 out LOUD. When I point to the 4th gummy bear, I will whisper 4. When I point to the 5th, I will whisper 5. When I point to the 6th*

gummy bear, I will say 6 out LOUD. Watch me count!

1 (whisper), *2* (whisper), *3* (LOUD), *4* (whisper), *5* (whisper), *6* (LOUD), *7* (whisper), *8* (whisper), *9* (LOUD), *10* (whisper), *11* (whisper), *12* (LOUD), *13* (whisper), *14* (whisper), *15* (LOUD), *16* (whisper), *17* (whisper), *18* (LOUD), *19* (whisper), *20* (whisper), *21* (LOUD), *22* (whisper), *23* (whisper), *24* (LOUD), *25* (whisper), *26* (whisper), *27* (LOUD), *28* (whisper), *29* (whisper), *30* (LOUD), *31* (whisper), *32* (whisper), *33* (LOUD), *34* (whisper), *35* (whisper), *36* (LOUD)

Parent: *Now we'll count these 36 gummy bears together. Let's use our counting fingers and our secret counting trick. (Shh). Get ready. Count: 1, 2, 3…36. Terrific counting!*

Parent: *Now it is your turn to count using our secret counting trick! Get ready! Count!*

Child counts to 36 using to whisper/loud technique and his/her counting fingers. Remember that it is common for children to mix up the volume in this activity. For example, he/she may whisper the 3 when he/she intends to say it loud or he/she may say the 4 loud when he/she intends to whisper it. Allow your child to self-correct his/her error rather than correcting him/her. Reassure your child and laugh with him/her over the error. Model the activity again if your child is unaware of his/her error.

Parent: *Now it is time to do some pages in the workbook.*

Child completes Lesson 5 in the workbook.

Parent: *Fantastic counting! There is just one more thing to do before we eat a few of our counters! We will recite the skip counting chant!*

Recite the skip counting chant using some type of physical movement. Remember to move around and be silly while reciting. Include any siblings or friends.

Parent: *Wonderful work! Now let's have a treat.*

Lesson 6

In This Lesson You Will:

1. Count with your child to 36 while stepping/clapping
2. Count with your child to 36 while jumping/clapping
3. Count with your child to 36 while catching
4. Have your child place 3 objects in each of the 12 compartments
5. Model counting objects by 1 using the "whisper/loud" counting trick
6. Count objects together by 1 using the "whisper/loud" counting trick
7. Have your child count objects by 1 using the "whisper/loud" counting trick
8. Model skip counting by 3's
9. Skip count together by 3's
10. Play the Sleeping Bears game
11. Have your child skip count by 3's
12. Play the Add On 3 game
13. Have your child complete the workbook lesson
14. Recite or sing the skip counting chant together with any siblings or friends.

Parent: *Let's start with counting our steps. We will count to 36 while stepping and clapping. Ready? Count! 1, 2, 3...36.*

Parent: *Now we will count with jumping instead of stepping. We will count to 36 while jumping and clapping. Ready? Jump: 1, 2, 3...36.*

Parent: *Now we will count with catching instead of jumping. We will count to 36 while we toss and catch this beanbag. We will count as we catch the beanbag. Get ready. Here it comes: 1, 2, 3...36.*

Parent: *Now it is time to count the gummy bears. Put 3 gummy bears into each compartment.*

Parent: *Now it is my turn to count the gummy bears using our secret counting method. Shh. I need my counting fingers. Have you seen my counting fingers? Where are they? Oh, here they are. Now I will count: 1, 2, 3...36.*

Parent: *Now we'll count these 36 gummy bears together, using our secret counting trick. (Shh). Are your counting fingers ready? Get ready. Count: 1, 2, 3...36.*

Help your child to find his/her counting fingers. Use the whisper/loud counting technique.

Parent: *Now it is your turn to count these 36 gummy bears. Use our secret counting trick and your counting fingers. Get ready. Count.*

Child counts to 36 using the whisper/loud technique and his/her counting fingers.

Parent: *Now I will skip count the gummy bears by 3's. This is called skip counting by 3's because I will skip all the numbers that I whispered when we counted with our secret counting trick. Skip counting sounds the same as our secret counting trick but it can be much, much faster. Watch how fast I can skip count these gummy bears. 3, 6, 9, 12, 15, 18, 21, 24, 27, 30, 33, 36.*

Be sure to point to the groups of 3 when you model skip counting. This helps your child to understand that when he/she skip counts, he/she is adding on 3 at a time.

Parent: *Now we'll skip count by 3's together. Are your counting fingers ready? Get ready. Count: 3, 6, 9, 12, 15, 18, 21, 24, 27, 30, 33, 36.*

If your child can count by 3's easily then skip the Sleeping Bears Game.

THE SLEEPING BEARS GAME

Dump out counting container.

Parent: *Now we will play a game called Sleeping Bears. We each take turns being the Mommy Bear counting our sleeping bear cubs.*

When we put the bear cubs in the compartment, we will pretend that we are tucking them in bed. We will tuck 3 bears in bed at a time. They fall asleep immediately! Then the Mommy needs to know how many of her little bear cubs are asleep so she counts them all up. You can be the Mommy Bear first.

Put these gummy bears in the first compartment. How many are there?

Have your child take 3 gummy bears; put them in the first compartment and say, "3 sleeping bear cubs."
Parent puts 3 gummy bears in the next available compartment and says, "3, 6 sleeping bear cubs."

Now your turn:
Child puts 3 gummy bears in the next available compartment and says, "3, 6, 9 sleeping bear cubs."
Parent puts 3 gummy bears in the next available compartment and says, "3, 6, 9, 12 sleeping bear cubs."
Continue until all compartments are filled. If your child likes the game enough, play it again. Be sure that 3 gummy bears are in each compartment.

Parent: *Now it is your turn to count by 3's. Get ready. Count.*

Child skip counts by 3's from 3 to 36, pointing with his/her counting fingers to each compartment of 3 gummy bears.

If your child has difficulty counting by 3's, count together and review this lesson.

Parent: *Excellent counting!*

THE ADD ON 3 GAME

Parent: *Now we will play the Add On 3 game. Did you know that when we skip count by 3's, we start at zero and add on 3? Then we keep adding on 3 to the previous number. Watch me add up all the gummy bears by adding on 3.*

Begin by pointing to an empty compartment to explain 0 or no gummy bears. Then point to each compartment as you add on the 3 gummy bears.

Parent: *0 or no gummy bears plus 3 gummy bears is equal to 3 gummy bears.*
3 gummy bears plus 3 gummy bears is equal to 6 gummy bears.
6 gummy bears plus 3 gummy bears is equal to 9 gummy bears.
9 gummy bears plus 3 gummy bears is equal to 12 gummy bears
12 gummy bears plus 3 gummy bears is equal to 15 gummy bears
15 gummy bears plus 3 gummy bears is equal to 18 gummy bears
18 gummy bears plus 3 gummy bears is equal to 21 gummy bears
21 gummy bears plus 3 gummy bears is equal to 24 gummy bears
24 gummy bears plus 3 gummy bears is equal to 27 gummy bears
27 gummy bears plus 3 gummy bears is equal to 30 gummy bears
30 gummy bears plus 3 gummy bears is equal to 33 gummy bears
33 gummy bears plus 3 gummy bears is equal to 36 gummy bears

Parent: *Now we'll skip count by 3's together by adding on 3.*

Repeat previous step.

Parent: *Now it is your turn to skip count by 3's by adding on 3.*

Repeat previous step.

Parent: *Now it is time to do some pages in the workbook.*

Child completes Lesson 6 in the workbook.

Parent: *There is just one more thing to do before we eat a few of our counters! We will recite the skip counting chant!*

Recite the skip counting chant using some type of physical movement. Remember to move around and be silly while reciting. Include any siblings or friends.

Parent: *Wonderful work. Now let's have a treat!*

Lesson 7

In This Lesson You Will:

1. Count with your child to 36 while stepping/clapping
2. Count with your child to 36 while jumping/clapping
3. Count with your child to 36 while catching
4. Have your child place 3 objects in each of the 12 compartments
5. Model counting objects by 1 using the "whisper/loud" counting trick
6. Count objects together by 1 using the "whisper/loud" counting trick
7. Have your child count objects by 1 using the "whisper/loud" counting trick
8. Model skip counting by 3's
9. Skip count together by 3's
10. Have your child skip counts by 3's
11. Play the Add On 3 game
12. Play the How Many? game
13. Have your child complete the workbook lesson
14. Recite or sing the skip counting chant together with any siblings or friends

Parent: *Let's start with counting our steps. We will count to 36 while stepping and clapping. Ready? Count! 1, 2, 3...36.*

Parent: *Now we will count with jumping instead of stepping. We will count to 36 while jumping and clapping. Ready? Jump. 1, 2, 3...36.*

Parent: *Now we will count with catching instead of jumping. We will count to 36 while we toss and catch this beanbag. We will count as we catch the beanbag. Get ready. Here it comes. 1, 2, 3...36.*

Parent: *Now it is time to count the gummy bears. Put 3 gummy bears into each compartment.*

Parent: *Now it is my turn to count the gummy bears using our secret counting method. Have you seen my counting fingers? Where are they? Oh, here they are. Now I will count: 1, 2, 3...36.*

Parent: *Now we'll count these 36 gummy bears together, using our secret counting trick. (Shh). Are your counting fingers ready? Get ready. Count: 1, 2, 3...36.*

Use the whisper/loud counting technique.

Parent: *Now it is your turn to count these 36 gummy bears. Use our secret counting trick and your counting fingers. Get ready. Count.*

Child counts to 36 using to whisper/loud technique and his/her counting fingers.

Parent: *Now I will skip count the gummy bears by 3's. Remember that this is called skip counting by 3's because I will skip all the numbers that I whispered when we counted with our secret counting trick. Listen to how fast I can skip count these gummy bears! 3, 6, 9, 12, 15, 18, 21, 24, 27, 30, 33, 36.*

Parent: *Now we'll skip count by 3's together. Do you think that you can be fast like me? Are your counting fingers ready? Get ready. Count: 3, 6, 9, 12, 15, 18, 21, 24, 27, 30, 33, 36.*

Have your child count along with you.

Parent: *Now it is your turn to count by 3's. Get ready. Count.*

Child skip counts by 3's from 3-36 pointing with counting fingers to each compartment of 3 gummy bears. Continue to practice this until child can easily count by 3's.

Parent: *Excellent counting!*

THE ADD ON 3 GAME

Parent: *Now we will play the Add On 3 game. Do you remember that when we skip count by 3's, we start at zero and add on 3? Then we keep adding on 3 to the previous number. Watch me add* *up all the gummy bears by adding on 3.*

Begin by pointing to an empty compartment to explain 0 or no gummy bears. Then point to each compartment as you add on the 3 gummy bears.

Parent: *0 or no gummy bears plus 3 gummy bears is equal to 3 gummy bears.*
3 gummy bears plus 3 gummy bears is equal to 6 gummy bears.
6 gummy bears plus 3 gummy bears is equal to 9 gummy bears.
9 gummy bears plus 3 gummy bears is equal to 12 gummy bears
12 gummy bears plus 3 gummy bears is equal to 15 gummy bears
15 gummy bears plus 3 gummy bears is equal to 18 gummy bears
18 gummy bears plus 3 gummy bears is equal to 21 gummy bears
21 gummy bears plus 3 gummy bears is equal to 24 gummy bears
24 gummy bears plus 3 gummy bears is equal to 27 gummy bears
27 gummy bears plus 3 gummy bears is equal to 30 gummy bears
30 gummy bears plus 3 gummy bears is equal to 33 gummy bears
33 gummy bears plus 3 gummy bears is equal to 36 gummy bears

Parent: *Now we'll skip count by 3's together by adding on 3.*

Repeat previous step.

Parent: *Now it is your turn to skip count by 3's by adding on 3.*

Repeat previous step.

THE HOW MANY? GAME

Parent: *Now it is time to play the "How Many?" game. Put 3 gummy bears in the first compartment.*

Dump out counting container and fill one compartment with 3 gummy bears.

Parent: *How many compartments have gummy bears?*

Child should answer that 1 compartment is filled.

Parent: *How many gummy bears are there?*

Child should answer that there are 3 gummy bears.

Parent: *You just multiplied! 1 times 3 is 3. 1 compartment filled with 3 gummy bears is 3 gummy bears. Now fill up another compartment with 3 gummy bears.*

Child puts 3 gummy bears in another compartment.

Parent: *How many compartments have gummy bears?*

Child should answer that 2 compartments are filled.

Parent: *How many gummy bears are there?*

Child should skip count and determine that there are 6 total gummy bears.

Parent: *You just multiplied again! 2 times 3 is 6. 2 compartments filled with 3 gummy bears are 6 gummy bears.*

Parent: *This is written as 2×3.*

Show your child on paper how 2 x 3 looks. Then show him/her that multiplication is also written as one number over the other with the multiplication or times symbol, ×, written next to the bottom number. For example,

$$\begin{array}{r} 2 \\ \times 3 \\ \hline \end{array}$$

Parent: *Now fill up another compartment with 3 gummy bears. How many compartments have gummy bears?*

Child should answer that 3 compartments are filled.

Parent: *How many gummy bears are there?*

Child should skip count and determine that there are 9 total gummy bears.

Parent: *You multiplied again! 3 times 3 is 9. 3 compartments filled with 3 gummy bears are 9 gummy bears. This is written as 3×3.*

Show your child on paper how 3 x 3 looks.

Parent: *Now fill up another compartment with 3 gummy bears. How many compartments have gummy bears?*

Child should answer that 4 compartments are filled.

Parent: *How many gummy bears are there?*

Child should skip count and determine that there are 12 total gummy bears.

Parent: *Terrific multiplying! 4 times 3 is 12. 4 compartments filled with 3 gummy bears are 12 gummy bears. This is written as 4×3.*

Show your child on paper how 4 x 3 looks.

Parent: *Now fill up another compartment with 3 gummy bears. How many compartments have gummy bears?*

Child should answer that 5 compartments are filled.

Parent: *How many gummy bears are there?*

Child should skip count and determine that there are 15 total gummy bears.

Parent: *You multiplied again! 5 times 3 is 15. 5 compartments filled with 3 gummy bears are 15 gummy bears. This is written as 5×3.*

Show your child on paper how 5 x 3 looks.

Parent: *Now fill up another compartment with 3 gummy bears. How many compartments have gummy bears?*

Child should answer that 6 compartments are filled.

Parent: *How many gummy bears are there?*

Child should skip count and determine that there are 18 total gummy bears.

Parent: *Magnificent multiplying! 6 times 3 is 18. 6 compartments filled with 3 gummy bears are 18 gummy bears. This is written as 6×3.*

Show your child on paper how 6 x 3 looks.

Parent: *Now fill up another compartment with 3 gummy bears. How many compartments have gummy bears?*

Child should answer that 7 compartments are filled.

Parent: *How many gummy bears are there?*

Child should skip count and determine that there are 21 total gummy bears.

Parent: *You just multiplied again! 7 times 3 is 21. 7 compartments filled with 3 gummy bears are 21 gummy bears. This is written as 7×3.*

Show your child on paper how 7 x 3 looks.

Parent: *Now fill up another compartment with 3 gummy bears. How many compartments have gummy bears?*

Child should answer that 8 compartments are filled.

Parent: *How many gummy bears are there?*

Child should skip count and determine that there are 24 total gummy bears.

Parent: *I bet you didn't know that you could multiply so well! But you did it! 8 times 3 is 24. 8 compartments filled with 3 gummy bears are 24 gummy bears. This is written as 8×3.*

Show your child on paper how 8 x 3 looks.

Parent: *Now fill up another compartment with 3 gummy bears. How many compartments have gummy bears?*

Child should answer that 9 compartments are filled.

Parent: *How many gummy bears are there?*

Child should skip count and determine that there are 27 total gummy bears.

Parent: *Yes. Your multiplication is correct. 9 times 3 is 27. 9 compartments filled with 3 gummy bears are 27 gummy bears. This is written as 9×3.*

Show your child on paper how 9 x 3 looks.

Parent: *Now fill up another compartment with 3 gummy bears. How many compartments have gummy bears?*

Child should answer that 10 compartments are filled.

Parent: *How many gummy bears are there?*

Child should skip count and determine that there are 30 total gummy bears.

Parent: *You multiplied correctly again! 10 times 3 is 30. 10 compartments filled with 3 gummy bears are 30 gummy bears. This is written as 10×3.*

Show your child on paper how 10 x 3 looks.

Parent: *I am very impressed with your stupendous multiplying Now it is time to do some pages in the workbook.*

Child completes Lesson 7 in the workbook.

Parent: *There is just one more thing to do before we eat a few of our counters! We will recite the skip counting chant!*

Recite the skip counting chant using some type of physical movement. Remember to move around and be silly while reciting. Include any siblings or friends.

Parent: *Wonderful work. Now let's have a treat!*

Lesson 8

In This Lesson You Will:

1. Count with your child to 36 while stepping/clapping
2. Count with your child to 36 while jumping/clapping
3. Count with your child to 36 while catching
4. Have your child place 3 objects in each of the 12 compartments
5. Model counting objects by 1 using the "whisper/loud" counting trick
6. Count objects together by 1 using the "whisper/loud" counting trick
7. Have your child count objects by 1 using the "whisper/loud" counting trick
8. Model skip counting by 3's
9. Skip count together by 3's
10. Have your child skip count by 3's
11. Play the How Many? game
12. Play the Finger Counting game
13. Model using fingers to solve multiplication problems
14. Practice solving multiplication problems together
15. Have your child solve multiplication problems
16. Have your child complete the workbook lesson
17. Recite or sing the skip counting chant together with any siblings or friends

Parent: *Let's start with counting our steps. We will count to 36 while stepping and clapping. Ready? Count! 1, 2, 3...36.*

Parent: *Now we will count with jumping instead of stepping. We will count to 36 while jumping and clapping. Ready? Jump. 1, 2, 3...36.*

Parent: *Now we will count with catching instead of jumping. We will count to 36 while we toss and catch this beanbag. We will count as we catch the beanbag. Get ready. Here it comes. 1, 2, 3...36.*

Parent: *Now it is time to count the gummy bears. Put 3 gummy bears into each compartment.*

Parent: *Now it is my turn to count the gummy bears using our secret counting method. 1, 2, 3...36.*

Parent: *Now we'll count these 36 gummy bears together, with our secret counting trick. (Shh). Show me your counting fingers! Get ready. Count: 1, 2, 3...36.*

Parent and child count to 36 using the whisper/loud counting technique.

Parent: *Now it is your turn to count these 36 gummy bears with our secret counting trick and your counting fingers. Get ready. Count.*

Child counts to 36 using the whisper/loud counting technique and his/her counting fingers.

Parent: *Now I will skip count the gummy bears by 3's. Listen to how fast I can skip count these gummy bears! 3, 6, 9, 12, 15, 18, 21, 24, 27, 30, 33, 36.*

Parent: *Now we'll skip count by 3's together. Do you think that you can count fast like me? Are your counting fingers ready? Get ready. Count: 3, 6, 9, 12, 15, 18, 21, 24, 27, 30, 33, 36.*

Have your child count along with you. Be sure that your child points to each compartment as he/she skip counts.

Parent: *Now it is your turn to count by 3's. Get ready. Count.*

Child skip counts by 3's from 3 to 36, pointing with counting fingers to each compartment of 3 gummy bears. Continue to practice this until your child can easily count by 3's.

Parent: *Excellent counting!*

THE HOW MANY? GAME

Parent: *Now it is time to play the "How Many?" game. Put 3 gummy bears in the first compartment.*

Dump out counting container and fill one compartment with 3 gummy bears.

Parent: *How many compartments have gummy bears?*

Child should answer that 1 compartment is filled.

Parent: *How many gummy bears are there?*

Child should answer that there are 3 gummy bears.

Parent: *You just multiplied! 1 times 3 is 3. 1 compartment filled with 3 gummy bears is 3 gummy bears. Now fill up another compartment with 3 gummy bears.*

Child puts 3 gummy bears in another compartment.

Parent: *How many compartments have gummy bears?*

Child should answer that 2 compartments are filled.

Parent: *How many gummy bears are there?*

Child should skip count and determine that there are 6 total gummy bears.

Parent: *You just multiplied again! 2 times 3 is 6. 2 compartments filled with 3 gummy bears are 6 gummy bears.*

Parent: *This is written as 2×3.*

Show your child on paper how 2 x 3 looks. Then show him/her that multiplication is also written as one number over the other with the multiplication or times symbol, ×, written next to the bottom number. For example,

$$\begin{array}{r} 2 \\ \times 3 \\ \hline \end{array}$$

Parent: *Now fill up another compartment with 3 gummy bears. How many compartments have gummy bears?*

Child should answer that 3 compartments are filled.

Parent: *How many gummy bears are there?*

Child should skip count and determine that there are 9 total gummy bears.

Parent: *You multiplied again! 3 times 3 is 9. 3 compartments filled with 3 gummy bears are 9 gummy bears. This is written as 3×3.*

Show your child on paper how 3 x 3 looks.

Parent: *Now fill up another compartment with 3 gummy bears. How many compartments have gummy bears?*

Child should answer that 4 compartments are filled.

Parent: *How many gummy bears are there?*

Child should skip count and determine that there are 12 total gummy bears.

Parent: *Terrific multiplying! 4 times 3 is 12. 4 compartments filled with 3 gummy bears are 12 gummy bears. This is written as 4×3.*

Show your child on paper how 4 x 3 looks.

Parent: *Now fill up another compartment with 3 gummy bears. How many compartments have gummy bears?*

Child should answer that 5 compartments are filled.

Parent: *How many gummy bears are there?*

Child should skip count and determine that there are 15 total gummy bears.

Parent: *You multiplied again! 5 times 3 is 15. 5 compartments filled with 3 gummy bears are 15 gummy bears. This is written as 5×3.*

Show your child on paper how 5 x 3 looks.

Parent: *Now fill up another compartment with 3 gummy bears. How many compartments have gummy bears?*

Child should answer that 6 compartments are filled.

Parent: *How many gummy bears are there?*

Child should skip count and determine that there are 18 total gummy bears.

Parent: *Splendid multiplying! 6 times 3 is 18. 6 compartments filled with 3 gummy bears are 18 gummy bears. This is written as 6×3.*

Show your child on paper how 6 x 3 looks.

Parent: *Now fill up another compartment with 3 gummy bears. How many compartments have gummy bears?*

Child should answer that 7 compartments are filled.

Parent: *How many gummy bears are there?*

Child should skip count and determine that there are 21 total gummy bears.

Parent: *You just multiplied again! 7 times 3 is 21. 7 compartments filled with 3 gummy bears are 21 gummy bears. This is written as 7×3.*

Show your child on paper how 7 x 3 looks.

Parent: *Now fill up another compartment with 3 gummy bears. How many compartments have gummy bears?*

Child should answer that 8 compartments are filled.

Parent: *How many gummy bears are there?*

Child should skip count and determine that there are 24 total gummy bears.

Parent: *I bet you didn't know that you could multiply so well! But you did it! 8 times 3 is 24. 8 compartments filled with 3 gummy bears are 24 gummy bears. This is written as 8×3.*

Show your child on paper how 8 x 3 looks.

Parent: *Now fill up another compartment with 3 gummy bears. How many compartments have gummy bears?*

Child should answer that 9 compartments are filled.

Parent: *How many gummy bears are there?*

Child should skip count and determine that there are 27 total gummy bears.

Parent: *Yes. Your multiplication is correct. 9 times 3 is 27. 9 compartments filled with 3 gummy bears are 27 gummy bears. This is written as 9×3.*

Show your child on paper how 9 x 3 looks.

Parent: *Now fill up another compartment with 3 gummy bears. How many compartments have gummy bears?*

Child should answer that 10 compartments are filled.

Parent: *How many gummy bears are there?*

Child should skip count and determine that there are 30 total gummy bears.

Parent: *You multiplied correctly again! 10 times 3 is 30. 10 compartments filled with 3 gummy bears are 30 gummy bears. This is written as 10×3.*

Show your child on paper how 10 x 3 looks.

Parent: *I am very impressed with your superb multiplying*

THE FINGER COUNTING GAME

Parent: *Now it is time to play a game called Finger Counting. Let's pretend I have 3 dots on each of my fingers. When I skip count, I will be counting the pretend dots on my fingers.*

Skip count on your fingers by gently squeezing the top of your finger with the thumb and index finger of your other hand. Start with the pinky on your left hand.
Squeeze the top of your pinky and say 3.
Squeeze the top of your ring finger and say 6.
Squeeze the top of your middle finger and say 9.
Squeeze the top of your index finger and say 12.
Squeeze the top of your thumb and say 15.
Switch hands and squeeze the top of your right thumb while saying 18. Count 21, 24, 27 and 30 on your right index, middle, ring finger and pinky, consecutively. Borrow 2 fingers from your child and count 33 and 36.

Parent: *Now it is your turn to play Finger Counting. Wiggle your fingers. Skip count on your fingers.*

Help your child to skip count by 3's on his/her fingers. Borrow 2 fingers from you for 33 and 36. Practice 3 times or until your child is comfortable with Finger Counting.

Parent: *Do you remember how great fingers are for solving multiplication problems? I will show you how to use fingers to multiply 3×3.*

I hold up 3 fingers and skip count 3 times by 3's, 3, 6, 9. Remember that we pretend that there are 3 dots on each of my 3 fingers. The answer to 3×3 is 9.

Hold up 3 fingers and emphasize squeezing the top of each finger while counting 3, 6, 9.

Parent: *We don't need to put dots on our fingers to figure out multiplication problems. We can pretend that we have dots on our fingers whenever we skip count on our fingers, to help us to figure* *out multiplication problems.*

Parent: *Now we'll figure out 4×3 together. Hold up 4 fingers and skip count by 3's: 3, 6, 9, 12. So the answer to 4×3 is 12.*

Help your child to hold up 4 fingers and use his/her fingers to skip count: 3, 6, 9, 12.

Parent: *Now it's your turn to figure out 3×3. Hold up 3 fingers and skip count.*

Help your child to hold up 3 fingers and skip count. Child should answer that 3 times 3 is 9.

Parent: *Now it's your turn to figure out 4×3. Hold up 4 fingers and skip count.*

Help your child to hold up 4 fingers and skip count. Child should answer that 4 times 3 is 12.

Parent: *What about 5×3? Hold up 5 fingers and skip count.*

Child holds up 5 fingers and skip counts. Child should answer that 5 times 3 is 15.

Parent: *What about 6×3? Hold up 6 fingers and skip count.*

Child holds up 6 fingers and skip counts. Child should answer that 6 times 3 is 18.

Parent: *What about 7×3? Hold up 7 fingers and skip count.*

Child holds up 7 fingers and skip counts. Child should answer that 7 times 3 is 21.

Parent: *What about 8×3? Hold up 8 fingers and skip count.*

Child holds up 8 fingers and skip counts. Child should answer that 8 times 3 is 24.

Parent: *What about 9×3? Hold up 9 fingers and skip count.*

Child holds up 9 fingers and skip counts. Child should answer that 9 times 3 is 27.

Parent: *What about 10×3? Hold up 10 fingers and skip count.*

Child holds up 10 fingers and skip counts. Child should answer that 10 times 3 is 30.

Parent: *Now you know how to multiply by 3's! Let's do some pages in the workbook.*

Child completes Lesson 8 in the workbook.

Parent: *There is just one more thing to do before we eat a few of our counters! We will recite the skip counting chant!*

Recite the skip counting chant using some type of physical movement. Remember to move around and be silly while reciting. Include any siblings or friends.

Parent: *Wonderful work. Now let's have a treat!*

IMPORTANT NOTE:

Congratulations. Now that you have completed lessons 1 to 8, you are well acquainted with the mechanics of the program. In lessons 1 to 8, the detailed outline of the lesson and the complete script is provided. The instructions for each game and activity are described in detail.

In lessons 9 to 45, only the detailed outline of the lesson is included. (When new material is introduced, the complete script will again be provided.) If you need a refresher on how to play a game or do an activity, refer back to an earlier script. The detailed outline in each lesson lists the page number for scripted instructions. Go to the page number and substitute the appropriate number(s) into the script. For example, in lesson 9, step 1 is to count to 48 while stepping and clapping. Refer back to lesson 1 for the complete script, and substitute 48 into the instructions. Count to 48 (rather than 24) while stepping and clapping.

Lesson 9

In This Lesson You Will:

1. Count with your child to 48 while stepping/clapping
2. Count with your child to 48 while playing catch
3. Have your child place 4 objects in each of the 12 compartments
4. Model counting objects by 1 with "counting fingers"
5. Count objects together by 1 with "counting fingers"
6. Have your child count objects by 1 with "counting fingers"
7. Model skip counting by 4's
8. Model the "whisper/loud" counting trick
9. Count objects together using the "whisper/loud" counting trick
10. Have your child count objects using the "whisper/loud" counting trick
11. Have your child skip count by 2's and 3's on fingers and review (see below)
12. Have your child complete the workbook lesson
13. Recite or sing the skip counting chant together with any siblings or friends.

Parent: *Now it is time to review. Wiggle your fingers and skip count by 2's, then 3's on your fingers.*

Child skip counts using his/her fingers. Use 2 of your fingers for the 11th and 12th fingers.

Parent: *What is 4×2?*

Child holds up 4 fingers and skip counts. Child should answer that 4 times 2 is 8.

Parent: *What is 4×3?*

Child holds up 4 fingers and skip counts. Child should answer that 4 times 3 is 12.

Lesson 10

In This Lesson You Will:

1. Count with your child to 48 while stepping/clapping
2. Count with your child to 48 while catching
3. Have your child places 4 objects in each of the 12 compartments
4. Model counting objects by 1 using the "whisper/loud" counting trick
5. Count objects together by 1 using the "whisper/loud" counting trick
6. Have your child count objects by 1 using the "whisper/loud" counting trick
7. Model skip counting by 4s
8. Skip count together by 4's
9. Play the Sleeping Bears game; substitute 4 into script on page 38
10. Have your child skip count by 4's
11. Play the Add On 4 game; substitute 4 into script on page 39
12. Have your child skip count by 2's and 3's on fingers and review (see below)
13. Have your child complete the workbook lesson
14. Recite or sing the skip counting chant together with any siblings or friends

Parent: *Now it is time to review. Wiggle your fingers and skip count by 2's, then 3's on your fingers.*

Child skip counts using his/her fingers. Use 2 of your fingers for the 11th and 12th fingers.

Parent: *What is 6×2?*

Child holds up 6 fingers and skip counts. Child should answer that 6 times 2 is 12.

Lesson 11

In This Lesson You Will:

1. Count with your child to 48 while stepping/clapping
2. Count with your child to 48 while catching
3. Have your child place 4 objects in each of the 12 compartments
4. Model counting objects by 1 using the "whisper/loud" counting trick
5. Count objects by 1 using the "whisper/loud" counting trick
6. Have your child count objects by 1 using the "whisper/loud" counting trick
7. Model skip counting by 4's
8. Skip count together by 4's
9. Have your child skip count by 4's
10. Play the Add On 4 game; substitute 4 into script on page 39
11. Play the How Many? game; substitute 4 into script on page 43
12. Have your child complete the workbook lesson
13. Have your child skip count by 2's and 3's on fingers and review (see below)
14. Recite or sing the skip counting chant together with any siblings or friends

Parent: *Now it is time to review. Wiggle your fingers and skip count by 2's, then 3's on your fingers.*

Child skip counts using his/her fingers. Use 2 of your fingers for the 11th and 12th fingers.

Parent: *What is 4×2?*

Child holds up 4 fingers and skip counts. Child should answer that 4 times 2 is 8.

Parent: *What is 2×3?*

Child holds up 2 fingers and skip counts. Child should answer that 2 times 3 is 6.

Lesson 12

In This Lesson You Will:

1. Count with your child to 48 while stepping/clapping
2. Count with your child to 48 while catching
3. Have your child place 4 objects in each of the 12 compartments
4. Model counting objects by 1 using the "whisper/loud" counting trick
5. Count objects together by 1 using the "whisper/loud" counting trick
6. Have your child count objects by 1 using the "whisper/loud" counting trick
7. Model skip counting by 4's
8. Skip count together by 4's
9. Have your child skip count by 4's
10. Play the How Many? game, substitute 4 into script on page 43
11. Play the Finger Counting game, substitute skip counting by 4, (4, 8, 12, 16, 20, 24, 28, 32, 36, 40, 44, 48) into script on page 55
12. Model using fingers to solve multiplication problems
13. Practice solving multiplication problems together
14. Have your child solve multiplication problems
15. Have your child skip count by 2's and 3's on fingers and review (see below)
16. Have your child complete the workbook lesson
17. Recite or sing the skip counting chant together with any siblings or friends

Parent: *Now it is time to review. Wiggle your fingers and skip count by 2's, then 3's on your fingers.*

Child skip counts using his/her fingers. Use 2 of your fingers for the 11th and 12th fingers.

Parent: *What is 5×2?*

Child holds up 5 fingers and skip counts. Child should answer that 5 times 2 is 10.

Lesson 13

In This Lesson You Will:

1. Count with your child to 60 while stepping/clapping
2. Count with your child to 60 while playing catch
3. Have your child place 5 objects in each of the 12 compartments
4. Model counting objects by 1 with "counting fingers"
5. Count objects together by 1 with "counting fingers"
6. Have your child count objects by 1 with "counting fingers"
7. Model skip counting by 5's
8. Model the "whisper/loud" counting trick
9. Count objects together using the "whisper/loud" counting trick
10. Have your child count objects using the "whisper/loud" counting trick
11. Have your child skip count by 2's, 3's and 4's on fingers and review (see below)
12. Have your child complete the workbook lesson
13. Recite or sing the skip counting chant together with any siblings or friends

Children find skip counting by 5's fun and easy. Encourage your child's enthusiasm by having friends and relatives ask your child to demonstrate skip counting by 5's.

Parent: *Now it is time to review. Wiggle your fingers and skip count by 2's, 3's, then 4's on your fingers.*

Child skip counts using his/her fingers. Use 2 of your fingers for the 11th and 12th fingers.

Parent: *What is 4×2?*

Child holds up 4 fingers and skip counts. Child should answer that 4 times 2 is 8.

Parent: *What is 4×3?*

Child holds up 4 fingers and skip counts. Child should answer that 4 times 3 is 12.

Lesson 14

In This Lesson You Will:

1. Count with your child to 60 while stepping/clapping
2. Count with your child to 60 while catching
3. Have your child place 5 objects in each of the 12 compartments
4. Model counting objects by 1 using the "whisper/loud" counting trick
5. Count objects together by 1 using the "whisper/loud" counting trick
6. Have your child count objects by 1 using the "whisper/loud" counting trick
7. Model skip counting by 5's
8. Skip count together by 5's
9. Play the Sleeping Bears game; substitute 5 into script on page 38
10. Have your child skip count by 5's
11. Play the Add On 5 game; substitute 5 into script on page 39
12. Have your child skip count by 2's, 3's and 4's on fingers and review (see below)
13. Have your child complete the workbook lesson
14. Recite or sing the skip counting chant together with any siblings or friends

Parent: *Now it is time to review. Wiggle your fingers and skip count by 2's, 3's, and then 4's on your fingers.*

Child skip counts using his/her fingers. Use 2 of your fingers for the 11th and 12th fingers.

Parent: *What is 4×4?*

Child holds up 4 fingers and skip counts. Child should answer that 4 times 4 is 16.

Parent: *What is 4×3?*

Child holds up 4 fingers and skip counts. Child should answer that 4 times 3 is 12.

Lesson 15

In This Lesson You Will:

1. Count with your child to 60 while stepping/clapping
2. Count with your child to 60 while catching
3. Have your child place 5 objects in each of the 12 compartments
4. Model counting objects by 1 using the "whisper/loud" counting trick
5. Count objects together by 1 using the "whisper/loud" counting trick
6. Have your child count objects by 1 using the "whisper/loud" counting trick
7. Model skip counting by 5's
8. Skip count together by 5's
9. Have your child skip count by 5's
10. Play the Add On 5 game; substitute 5 into script on page 39
11. Play the How Many? game; substitute 5 into script on page 43
12. Have your child skip count by 2's, 3's and 4's on fingers and review (see below)
13. Have your child complete the workbook lesson
14. Recite or sing the skip counting chant together with any siblings or friends

Parent: *Now it is time to review. Wiggle your fingers and skip count by 2's, 3's, and then 4's on your fingers.*

Child skip counts using his/her fingers. Use 2 of your fingers for the 11th and 12th fingers.

Parent: *What is 1×2?*

Child holds up 1 finger and skip counts. Child should answer that 1 times 2 is 2.

Parent: *What is 4×3?*

Child holds up 4 fingers and skip counts. Child should answer that 4 times 3 is 12.

Lesson 16

In This Lesson You Will:

1. Count with your child to 60 while stepping/clapping
2. Count with your child to 60 while catching
3. Have your child place 5 objects in each of the 12 compartments
4. Model counting objects by 1 using the "whisper/loud" counting trick
5. Count objects together by 1 using the "whisper/loud" counting trick
6. Have your child count objects by 1 using the "whisper/loud" counting trick
7. Model skip counting by 5's
8. Skip count together by 5's
9. Have your child skip count by 5's
10. Play the How Many? game; substitute 5 into script on page 43
11. Play the Finger Counting game; substitute 5 into script on page 55
12. Model using fingers to solve multiplication problems
13. Practice solving multiplication problems together
14. Have your child solve multiplication problems
15. Have your child skip count by 2's, 3's and 4's on fingers and review (see below)
16. Have your child complete the workbook lesson
17. Recite or sing the skip counting chant together with any siblings or friends

Parent: *Now it is time to review. Wiggle your fingers and skip count by 2's, 3's, and then 4's on your fingers.*

Child skip counts using his/her fingers. Use 2 of your fingers for the 11th and 12th fingers.

Parent: *What is 6×2?*

Child holds up 6 fingers and skip counts. Child should answer that 6 times 2 is 12.

Parent: *What is 6×3?*

Child holds up 6 fingers and skip counts. Child should answer that 6 times 3 is 18.

Lesson 17

In This Lesson You Will:

1. Count with your child to 72 while stepping/clapping
2. Count with your child to 72 while playing catch
3. Have your child place 6 objects in each of the 12 compartments
4. Model counting objects by 1 with "counting fingers"
5. Count objects together by 1 with "counting fingers"
6. Have your child count objects by 1 with "counting fingers"
7. Model skip counting by 6's
8. Model the "whisper/loud" counting trick
9. Count objects together using the "whisper/loud" counting trick
10. Have your child count objects using the "whisper/loud" counting trick
11. Have your child skip count by 2's, 3's, 4's and 5's on fingers and review (see below)
12. Have your child complete the workbook lesson
13. Recite or sing the skip counting chant together with any siblings or friends

In previous lessons, your child may have laid out the gummy bears in random order. In this lesson your child will lay out the gummy bears in an ordered pattern. The purpose of this is to help your child see that 12 groups of 6 is also 24 groups of 3.

Parent: *In this lesson, you will lay out the gummy bears in a pattern. In each compartment, you will lay out the six gummy bears in two rows. Put three gummy bears into each of the two rows.*

Child puts three gummy bears into the compartments in two rows.

Parent: *How many groups of six gummy bears do you see?*

Child answers that there are 12 groups of gummy bears.

Parent: *How many groups of three gummy bears do you see? Are 12 groups of six gummy bears the same as 24 groups of three gummy bears?*

Child answers that there are 24 groups of gummy bears. Yes. 12 groups of 6 gummy bears is the same as 24 groups of 3 gummy bears.

Parent: *Now it is time to review. Wiggle your fingers and then skip count by 2's, then 3's, then 4's and then 5's on our fingers.*

Child skip counts using fingers. Borrow 2 fingers from you for the 11th and 12th finger.

Parent: *What is 4×2?*

Child holds up 4 fingers and skip counts. Child should answer that 4 times 2 is 8.

Parent: *What is 4×3?*

Child holds up 4 fingers and skip counts. Child should answer that 4 times 3 is 12.

Parent: *What is 4×5?*

Child holds up 4 fingers and skip counts. Child should answer that 4 times 5 is 20.

Parent: *What is 5×2?*

Child holds up 5 fingers and skip counts. Child should answer that 5 times 2 is 10.

Parent: *What is 5×3?*

Child holds up 5 fingers and skip counts. Child should answer that 5 times 3 is 15.

Parent: *What is 8×2?*

Child holds up 8 fingers and skip counts. Child should answer that 8 times 2 is 16.

Parent: *What is 8×3?*

Child holds up 8 fingers and skip counts. Child should answer that 8 times 3 is 24.

Parent: *What is 8×4?*

Child holds up 8 fingers and skip counts. Child should answer that 8 times 4 is 32.

Lesson 18

In This Lesson You Will:

1. Count with your child to 72 while stepping/clapping
2. Count with your child to 72 while catching
3. Have your child place 6 objects in each of the 12 compartments
4. Model counting objects by 1 using the "whisper/loud" counting trick
5. Count objects together by 1 using the "whisper/loud" counting trick
6. Have your child count objects by 1 using the "whisper/loud" counting trick
7. Model skip counting by 6's
8. Skip count together by 6's
9. Play the Sleeping Bears game; substitute 6 into script on page 38
10. Play My Turn, Your Turn (see below)
11. Have your child skip count by 6's
12. Play the Add On 6 game; substitute 6 into script on page 39
13. Have your child skip count by 2's, 3's, 4's and 5's on fingers and review (see below)
14. Have your child complete the workbook lesson
15. Recite or sing the skip counting chant together with any siblings or friends

Parent: *Now we'll count by 6's together. Let's start with 6, 12, 18, 24, 30. Get ready! Count! 6, 12, 18, 24, 30.*

Stand up and practice this until your child really knows it, (minimum 6-10 times) before adding 36, 42, 48, 54, 60, 66 and 72. Clap and walk as you count. Add 36, 42, 48, 54, 60, 66 and 72, (additional multiples of 6), one at a time and then practice another 6 to 10 times if necessary. Your child may get the first 5 correct and then mix them all up when he/she adds the 6th. Take heart, he/she will learn this with practice and repetition just as long ago he/she learned to count from 1 to 10 by ones.

It is common for children to confuse skip counting by 6's with skip counting by 3's. Many of the numbers are the same. Confusion is minimized by repeating just the first five numbers of the skip counting pattern of 6's.

In this next game, it is important to continue only as long as your child gets most of the skip counting correct. Stop playing if your child is not able to concentrate. Try playing this game while making dinner, walking up the stairs, before bed, first thing in the morning or in the car. Each time you play, your child will remember a little bit more.

Learning to skip count by 6's takes extra time. Practice this game often. If this game is too difficult for your child, then continue to play only the first half of the game until your child is very comfortable with the first part of skip counting by 6's.

Parent: *Now we will play My Turn, Your Turn. Say 6.*

Child should say 6.

Parent: *6, 12. Now say 6, 12, 18.*

Child should say 6, 12, 18.

Parent: *6, 12, 18, 24. Now say 6, 12, 18, 24, 30.*

Child should say 6, 12, 18, 24, 30.

Parent: *6, 12, 18, 24, 30, 36. Now say 6, 12, 18, 24, 30, 36, 42.*

Child should say 6, 12, 18, 24, 30, 36, 42.

Parent: *6, 12, 18, 24, 30, 36, 42, 48. Now say 6, 12, 18, 24, 30, 36, 42, 48, 54.*

Child should say 6, 12, 18, 24, 30, 36, 42, 48, 54.

Parent: *6, 12, 18, 24, 30, 36, 42, 48, 54, 60. Now say 6, 12, 18, 24, 30, 36, 42, 48, 54, 60, 66.*

Child should say 6, 12, 18, 24, 30, 36, 42, 48, 54, 60, 66.

Parent: *Now it is time to review. Wiggle your fingers and then skip count by 2's, 3's, 4's, and 5's on your fingers.*

Child skip counts using fingers. Borrow 2 fingers from you for the 11th and 12th finger.

Parent: *What is 4×3?*

Child holds up 4 fingers and skip counts. Child should answer that 4 times 3 is 12.

Parent: *What is 4×4?*

Child holds up 4 fingers and skip counts. Child should answer that 4 times 4 is 16.

Lesson 19

In This Lesson You Will:

1. Count with your child to 72 while stepping/clapping
2. Count with your child to 72 while catching
3. Have your child place 6 objects in each of the 12 compartments
4. Model counting objects by 1 using the "whisper/loud" counting trick
5. Count objects together by 1 using the "whisper/loud" counting trick
6. Have your child count objects by 1 using the "whisper/loud" counting trick
7. Model skip counting by 6's
8. Skip count together by 6's
9. Play My Turn, Your Turn; see script on page 68
10. Have your child skip count by 6's
11. Play the Add On 6 game; substitute 6 into script on page 39
12. Play the How Many? game; substitute 6 into script on page 43
13. Have your child skip count by 2's, 3's, 4's and 5's on fingers and review (see below)
14. Have your child complete the workbook lesson
15. Recite or sing the skip counting chant together with any siblings or friends

Parent: *What is 7×2?*

Child holds up 7 fingers and skip counts. Child should answer that 7 times 2 is 14.

Parent: *What is 7×3?*

Child holds up 7 fingers and skip counts. Child should answer that 7 times 3 is 21.

Parent: *What is 7×4?*

Child holds up 7 fingers and skip counts. Child should answer that 7 times 4 is 28.

Lesson 20

In This Lesson You Will:

1. Count with your child to 72 while stepping/clapping
2. Count with your child to 72 while catching
3. Have your child place 6 objects in each of the 12 compartments
4. Model counting objects by 1 using the "whisper/loud" counting trick
5. Count objects together by 1 using the "whisper/loud" counting trick
6. Have your child count objects by 1 using the "whisper/loud" counting trick
7. Model skip counting by 6's
8. Skip count together by 6's
9. Play My Turn, Your Turn; see script on page 68
10. Have your child skip count by 6's
11. Play the How Many? game; substitute 6 into script on page 43
12. Play the Finger Counting game; substitute 6 into script on page 55
13. Model using fingers to solve multiplication problems
14. Practice solving multiplication problems together
15. Have your child solve multiplication problems
16. Have your child skip count by 2's, 3's, 4's and 5's on fingers and review (see below)
17. Have your child complete the workbook lesson
18. Recite or sing the skip counting chant with any siblings or friends

Parent: *What is 9×3?*

Child holds up 9 fingers and skip counts. Child should answer that 9 times 3 is 27.

Parent: *What is 9×4?*

Child holds up 9 fingers and skip counts. Child should answer that 9 times 4 is 36.

Parent: *What is 9×5?*

Child holds up 9 fingers and skip counts. Child should answer that 9 times 5 is 45.

Lesson 21

In This Lesson You Will:

1. Count with your child to 84 while stepping/clapping
2. Count with your child to 84 while playing catch
3. Have your child place 7 objects in each of the 12 compartments
4. Model counting objects by 1 with "counting fingers"
5. Count objects together by 1 with "counting fingers"
6. Have your child count objects by 1 with "counting fingers"
7. Model skip counting by 7's
8. Model the "whisper/loud" counting trick
9. Count objects together using the "whisper/loud" counting trick
10. Have your child count objects using the "whisper/loud" counting trick
11. Have your child skip count by 2's, 3's, 4's, 5's and 6's on fingers and review (see below)
12. Have your child complete the workbook lesson
13. Recite or sing the skip counting chant together with any siblings or friends

Parent: *What is 2×3?*

Child holds up 2 fingers and skip counts. Child should answer that 2 times 3 is 6.

Parent: *What is 6×4?*

Child holds up 6 fingers and skip counts. Child should answer that 6 times 4 is 24.

Parent: *What is 5×5?*

Child holds up 5 fingers and skip counts. Child should answer that 5 times 5 is 25.

Parent: *What is 6×6?*

Child holds up 6 fingers and skip counts. Child should answer that 6 times 6 is 36.

Lesson 22

In This Lesson You Will:

1. Count with your child to 84 while stepping/clapping
2. Count with your child to 84 while catching
3. Have your child place 7 objects in each of the 12 compartments
4. Model counting objects by 1 using the "whisper/loud" counting trick
5. Count objects together by 1 using the "whisper/loud" counting trick
6. Have your child count objects by 1 using the "whisper/loud" counting trick
7. Model skip counting by 7's
8. Skip count together by 7's
9. Play the Sleeping Bears game; substitute 7 into script on page 38
10. Play My Turn, Your Turn (see below)
11. Have your child skip count by 7's
12. Play the Add On 7 game; substitute 7 into script on page 39
13. Have your child skip count by 2's, 3's, 4's, 5's and 6's on fingers and review (see below)
14. Have your child complete the workbook lesson
15. Recite or sing the skip counting chant together with any siblings or friends

Parent: *Now we will play My Turn, Your Turn. Say 7.*

Child should say 7.

Parent: *7, 14. Now say 7, 14, 21.*

Child should say 7, 14, 21.

Parent: *7, 14, 21, 28. Now say 7, 14, 21, 28, 35.*

Child should say 7, 14, 21, 28, 35.

Parent: *7, 14, 21, 28, 35, 42. Now say 7, 14, 21, 28, 35, 42, 49.*

Child should say 7, 14, 21, 28, 35, 42, 49.

Parent: *7, 14, 21, 28, 35, 42, 49, 56. Now say 7, 14, 21, 28, 35, 42, 49, 56, 63.*

Child should say 7, 14, 21, 28, 35, 42, 49, 56, 63.

Parent: *7, 14, 21, 28, 35, 42, 49, 56, 63, 70. Now say 7, 14, 21, 28, 35, 42, 49, 56, 63, 70, 77.*

Child should say 7, 14, 21, 28, 35, 42, 49, 56, 63, 70, 77.

Parent: *Good job with that game!*

Parent: *Now it is time to review. Wiggle your fingers and then skip count by 2's, 3's, 4's, 5's, and 6's on your fingers.*

Child skip counts using his/her fingers. Use 2 of your fingers for the 11th and 12th fingers.

Parent: *Now it is time for some practice. What is 6×3?*

Child holds up 6 fingers and skip counts. Child should answer that 6 times 3 is 18.

Parent: *What is 6×4?*

Child holds up 6 fingers and skip counts. Child should answer that 6 times 4 is 24.

Parent: *What is 6×5?*

Child holds up 6 fingers and skip counts. Child should answer that 6 times 5 is 30.

Parent: *What is 6×6?*

Child holds up 6 fingers and skip counts. Child should answer that 6 times 6 is 36.

Lesson 23

In This Lesson You Will:

1. Count with your child to 84 while stepping/clapping
2. Count with your child to 84 while catching
3. Have your child place 7 objects in each of the 12 compartments
4. Model counting objects by 1 using the "whisper/loud" counting trick
5. Count objects together by 1 using the "whisper/loud" counting trick
6. Have your child count objects by 1 using the "whisper/loud" counting trick
7. Model skip counting by 7's
8. Skip count together by 7's
9. Play My Turn, Your Turn; see script on page 73
10. Have your child skip count by 7's
11. Play the Add On 7 game; substitute 7 into script on page 39
12. Play the How Many? game; substitute 7 into script on page 43
13. Have your child skip count by 2's, 3's, 4's, 5's and 6's on fingers and review (see below)
14. Have your child complete the workbook lesson
15. Recite or sing the skip counting chant together with any siblings or friends

Parent: *What is 4×3?*

Child holds up 4 fingers and skip counts. Child should answer that 4 times 3 is 12.

Parent: *What is 4×4?*

Child holds up 4 fingers and skip counts. Child should answer that 4 times 4 is 16.

Parent: *What is 4×5?*

Child holds up 4 fingers and skip counts. Child should answer that 4 times 5 is 20.

Parent: *What is 4×6?*

Child holds up 4 fingers and skip counts. Child should answer that 4 times 6 is 24.

Lesson 24

In This Lesson You Will:

1. Count with your child to 84 while stepping/clapping
2. Count with your child to 84 while catching
3. Have your child place 7 objects in each of the 12 compartments
4. Model counting objects by 1 using the "whisper/loud" counting trick
5. Count objects together by 1 using the "whisper/loud" counting trick
6. Have your child count objects by 1 using the "whisper/loud" counting trick
7. Model skip counting by 7's
8. Skip count together by 7's
9. Play My Turn, Your Turn; see script on page 73
10. Have your child skip count by 7's
11. Play the How Many? game; substitute 7 into script on page 43
12. Play the Finger Counting game; substitute 7 into script on page 55
13. Model using fingers to solve multiplication problems
14. Practice solving multiplication problems together
15. Have your child solve multiplication problems
16. Have your child skip count by 2's, 3's, 4's, 5's and 6's on fingers and review (see below)
17. Have your child complete the workbook lesson
18. Recite or sing the skip counting chant together with any siblings or friends

Lesson 25

In This Lesson You Will:

1. Count with your child to 96 while stepping/clapping
2. Count with your child to 96 while playing catch
3. Have your child place 8 objects in each of the 12 compartments
4. Model counting objects by 1 with "counting fingers"
5. Count objects together by 1 with "counting fingers"
6. Have your child count objects by 1 with "counting fingers"
7. Model skip counting by 8's
8. Model the "whisper/loud" counting trick
9. Count objects together using the "whisper/loud" counting trick
10. Have your child count objects using the "whisper/loud" counting trick
11. Have your child skip count by 2's, 3's, 4's, 5's, 6's and 7's on fingers and review
12. Have your child complete the workbook lesson
13. Recite or sing the skip counting chant together with any siblings or friends

Lesson 26

In This Lesson You Will:

1. Count with your child to 96 while stepping/clapping
2. Count with your child to 96 while catching
3. Have your child place 8 objects in each of the 12 compartments
4. Model counting objects by 1 using the "whisper/loud" counting trick
5. Count objects together by 1 using the "whisper/loud" counting trick
6. Have your child count objects by 1 using the "whisper/loud" counting trick
7. Model skip counting by 8's
8. Skip count together by 8's
9. Play the Sleeping Bears game; substitute 8 into script on page 38
10. Count by 8's together. (see below)
11. Play My Turn, Your Turn (see below)
12. Write down the skip counting by 8's numbers. (see below)
13. Play the Add On 8 game; substitute 8 into script on page 39
14. Have your child skip count by 2's, 3's, 4's, 5's, 6's and 7's on fingers and review
15. Have your child complete the workbook lesson
16. Recite or sing the skip counting chant together with any siblings or friends

Parent: *Now we'll count by 8's together. Let's start with 8, 16, 24, 32, 40. Get ready! Count! 8, 16, 24, 32, 40.*

Stand up and practice this until your child really knows it. Clap and walk as you count. It is common for children to confuse skip counting by 8's with skip counting by 4's. Many of the numbers are the same. Confusion is minimized by repeating just the first five numbers of the skip counting pattern of 8's.

In this next game, it is important to continue only as long as your child gets most of the skip counting correct. Try playing this game while walking the dog, driving in the car, or shopping at the grocery store. Always be on the lookout for groups of 8 to count.

Parent: *Now we will play My Turn, Your Turn. Say 8.*

Child should say 8.

Parent: *8, 16. Now say 8, 16, 24.*

Child should say 8, 16, 24.

Parent: *8, 16, 24, 32. Now say 8, 16, 24, 32, 40.*

Child should say 8, 16, 24, 32, 40.

Parent: *8, 16, 24, 32, 40, 48. Now say 8, 16, 24, 32, 40, 48, 56.*

Child should say 8, 16, 24, 32, 40, 48, 56.

Parent: *8, 16, 24, 32, 40, 48, 56, 64. Now say 8, 16, 24, 32, 40, 48, 56, 64, 72.*

Child should say 8, 16, 24, 32, 40, 48, 56, 64, 72

Parent: *8, 16, 24, 32, 40, 48, 56, 64, 72, 80. Now say 8, 16, 24, 32, 40, 48, 56, 64, 72, 80, 88.*

Child should say 8, 16, 24, 32, 40, 48, 56, 64, 72, 80, 88.

Parent: *Now we'll write down the skip counting pattern of 8's together. Let's look for a pattern. This pattern will help us remember how to skip count by 8's.*

Write down 8, 16, 24, 32, 40, 48, 56, 64, 72, 80, 88, 96 in a list. Line up the numbers so that each number is directly below the previous number. If your child understands the word digit, use the word. If not, the concept of digit is explained in lesson 33.

Parent: *Look! I see a pattern with the first three numbers, 08, 16 and 24. The first part of the number (the first digit) increases. It starts at 0, then it's 1, and then it's 2.*

Point to the numbers in the list as you explain.

Parent: *The second part of the number (the second digit) has a pattern, too. It decreases by 2. It starts at 8, then it's 6, then it's 4. Do all the numbers follow this pattern?*

Yes. In the skip counting by 8's pattern, the first digit increases by 1 and the second digit decreases by 2 until 40. Starting at 48 the pattern continues to 80.

Parent: *Let's see if we can write down the numbers from memory. Then let's practice skip counting by 8's.*

Take turns writing down and reciting the skip counting pattern of numbers from memory.

Lesson 27

In This Lesson You Will:

1. Count with your child to 96 while stepping/clapping
2. Count with your child to 96 while catching
3. Have your child place 8 objects in each of the 12 compartments
4. Model counting objects by 1 using the "whisper/loud" counting trick
5. Count objects together by 1 using the "whisper/loud" counting trick
6. Have your child count objects by 1 using the "whisper/loud" counting trick
7. Model skip counting by 8's
8. Skip count together by 8's
9. Play My Turn, Your Turn; see script on page 78
10. Have your child skip count by 8's
11. Play the Add On 8 game; substitute 8 into script on page 39
12. Play the How Many? game; substitute 8 into script on page 43
13. Have your child skip count by 2's, 3's, 4's, 5's, 6's and 7's on fingers and review
14. Have your child complete the workbook lesson
15. Recite or sing the skip counting chant with any siblings or friends

Lesson 28

In This Lesson You Will:

1. Count with your child to 96 while stepping/clapping
2. Count with your child to 96 while catching
3. Have your child place 8 objects in each of the 12 compartments
4. Model counting objects by 1 using the "whisper/loud" counting trick
5. Count objects together by 1 using the "whisper/loud" counting trick
6. Have your child count objects by 1 using the "whisper/loud" counting trick
7. Model skip counting by 8's
8. Skip count together by 8's
9. Play My Turn, Your Turn; see script on page 78
10. Have your child skip count by 8's
11. Play the How Many? game; substitute 8 into script on page 43
12. Play the Finger Counting game; substitute 8 into script on page 55
13. Model using fingers to solve multiplication problems
14. Practice solving multiplication problems together
15. Have your child solve multiplication problems
16. Have your child skip count by 2's, 3's, 4's, 5's, 6's and 7's on fingers and review
17. Have your child complete the workbook lesson
18. Recite or sing the skip counting chant together with any siblings or friends

Lesson 29

In This Lesson You Will:

1. Count with your child to 108 while stepping/clapping
2. Count with your child to 108 while playing catch
3. Have your child place 9 objects in each of the 12 compartments
4. Model counting objects by 1 with "counting fingers"
5. Count objects together by 1 with "counting fingers"
6. Have your child count objects by 1 with "counting fingers"
7. Model skip counting by 9's
8. Model the "whisper/loud" counting trick
9. Count objects together using the "whisper/loud" counting trick
10. Have your child count objects using the "whisper/loud" counting trick
11. Have your child skip count by 2's, 3's, 4's, 5's, 6's, 7's and 8's on fingers and review
12. Have your child complete the workbook lesson
13. Recite or sing the skip counting chant together with any siblings or friends

Lesson 30

In This Lesson You Will:

1. Count with your child to 108 while stepping/clapping
2. Count with your child to 108 while catching
3. Have your child place 9 objects in each of the 12 compartments
4. Model counting objects by 1 using the "whisper/loud" counting trick
5. Count objects together by 1 using the "whisper/loud" counting trick
6. Have your child count objects by 1 using the "whisper/loud" counting trick
7. Model skip counting by 9's
8. Skip count together by 9's
9. Play the Sleeping Bears game; substitute 9 into script on page 38
10. Play My Turn, Your Turn (see below)
11. Have your child skip count by 9's
12. Play the Add On 9 game; substitute 9 in script, page 39
13. Have your child skip count by 2's, 3's, 4's, 5's, 6's, 7's and 8's on fingers and review
14. Have your child complete the workbook lesson
15. Recite or sing the skip counting chant together with any siblings or friends

Parent: *Now we will play My Turn, Your Turn. Say 9.*

Child should say 9

Parent: *9, 18. Now say 9, 18, 27.*

Child should say 9, 18, 27.

Parent: *9, 18, 27, 36. Now say 9, 18, 27, 36, 45.*

Child should say 9, 18, 27, 36, 45.

Parent: *9, 18, 27, 36, 45, 54. Now say 9, 18, 27, 36, 45, 54, 63.*

Child should say 9, 18, 27, 36, 45, 54, 63.

Parent: *9, 18, 27, 36, 45, 54, 63, 72. Now say 9, 18, 27, 36, 45, 54, 63, 72, 81.*

Child should say 9, 18, 27, 36, 45, 54, 63, 72, 81.

Parent: *9, 18, 27, 36, 45, 54, 63, 72, 81, 90.*
Now say 9, 18, 27, 36, 45, 54, 63, 72, 81, 90, 99.

Child should say 9, 18, 27, 36, 45, 54, 63, 72, 81, 90, 99.

Lesson 31

In This Lesson You Will:

1. Count with your child to 108 while stepping/clapping
2. Count with your child to 108 while catching
3. Have your child place 9 objects in each of the 12 compartments
4. Model counting objects by 1 using the "whisper/loud" counting trick
5. Count objects together by 1 using the "whisper/loud" counting trick
6. Have your child count objects by 1 using the "whisper/loud" counting trick
7. Model skip counting by 9's
8. Skip count together by 9's
9. Play My Turn, Your Turn; see script on page 83
10. Have your child skip count by 9's
11. Play the Add On 9 game; substitute 9 into script on page 39
12. Play the How Many? game; substitute 9 into script on page 43
13. Have child skip count by 2's, 3's, 4's, 5's, 6's, 7's and 8's on fingers and review
14. Have your child complete the workbook lesson
15. Recite or sing the skip counting chant together with any siblings or friends

Lesson 32

In This Lesson You Will:

1. Count with your child to 108 while stepping/clapping
2. Count with your child to 108 while catching
3. Have your child place 9 objects in each of the 12 compartments
4. Model counting objects by 1 using the "whisper/loud" counting trick
5. Count objects together by 1 using the "whisper/loud" counting trick
6. Have your child count objects by 1 using the "whisper/loud" counting trick
7. Model skip counting by 9's
8. Skip count together by 9's
9. Play My Turn, Your Turn; see script on page 83
10. Have your child skip count by 9's
11. Play the How Many? game; substitute 9 into script on page 43
12. Play the Finger Counting game; substitute 9 into script on page 55
13. Model using fingers to solve multiplication problems
14. Practice solving multiplication problems
15. Have your child solve multiplication problems
16. Have your child skip count by 2's, 3's, 4's, 5's, 6's, 7's and 8's on fingers and review
17. Have your child complete the workbook lesson
18. Recite or sing the skip counting chant together with any siblings or friends

Parent: *Your fingers hold a fabulous trick for multiplying by 9's! Here is how to figure out the answer to 2x9:*

Hold both hands in front of you, with fingers stretched out.

Since you are multiplying 9 by 2, count two fingers from the left and bend down the second finger, which will be the ring finger on your left hand. Bend only the ring finger.

To the left of the bent finger is 1 finger, your left hand pinky. The first digit in your answer will be 1 since there is 1 to the left of the bent down finger.

Count the number of fingers to the right of the one that is bent down. In this case, there are 8 other fingers, 3 on your left hand and 5 on your right hand. (Count the 2 thumbs as fingers.) The second digit in your answer will be 8.

The answer is the combination of the two digits, 1 and 8. Combine the 1 that is on the left and the 8 on the right to form 18. The answer to 2x9 is 18.

Here is another example of the nines trick. Multiply 9x7

1. *Now let's multiply 9x7. Hold both hands in front of you, with fingers stretched out.*

2. *Since you are multiplying 9 by 7, count seven fingers from the left and bend down the seventh finger, which will be the index finger on your right hand. Bend only the index finger.*

3. *To the left of the bent finger is 6 fingers. The first digit in your answer will be 6.*

4. *Count the number of fingers to the right of the bent down finger. In this case, 3 fingers. The second digit in your answer will be 3.*

5. *The answer is the combination of the two digits, 6 and 3. Combine the 6 that is on the left and the 3 on the right to form 63. The answer to 9x7 is 63.*

Try out this fun trick for the other multiples of 9.

Lesson 33

In This Lesson You Will:

1. Count with your child to 120 while stepping/clapping
2. Count with your child to 120 while playing catch
3. Explain place value
4. Have your child skip count by 2's, 3's, 4's, 5's, 6's, 7's, 8's and 9's on fingers and review
5. Have your child complete the workbook lesson
6. Recite or sing the skip counting chant together with any siblings or friends

IMPORTANT NOTES:

This lesson assumes that your child can count by 10's. If your child does not know how to count by 10's, complete *only* the first part of lesson 34 (which covers skip counting by 10's) before proceeding.

Your child may grow bored counting large numbers by 1's. In this lesson and the following lessons, use money for counting rather than candy or whatever you have used in the past. The following script uses 12 dimes and 9 pennies.

Repeat this lesson or any part of this lesson, as many times as necessary, to ensure that your child solidly understands the material.

This is a complicated lesson. Remember to speak slowly.

Move ahead to lesson 34 if your child already understands place value and has mastered the concept.

Parent: *Did you know that the number 12 has 2 digits? The 2 digits are a 1 and a 2.*

Write down the number 12 on a piece of paper to show your child.

Parent: *Now I will write down the number 56. How many digits does the number 56 have? What is the first digit? What is the second digit?*

Child should answer that 56 has 2 digits. The first digit is 5 and the second digit is 6.

Parent: *Now I will write down the number 589. How many digits* *does the number 589 have? Name each of the digits.*

Child should answer that 589 has 3 digits. The digits are 5, 8 and 9 respectively. Continue practicing with naming digits until your child has mastered the concept.

Parent: *Now we will use dimes and pennies to build numbers. When I wrote down the number 12, it looked like this.*

Write down the number 12 again.

Parent: *This is what the number 12 looks like in dimes and pennies. Do you see that there is 1 dime and 2 pennies?*

Put 1 dime and 2 pennies on the table in front of you.

Parent: *Do you know what cents means?*

Child should answer 1 cent is the same as 1 penny.

Parent: *Do you see anything on the table that is the same as 10 pennies?*

Your child may know that a dime is the same as 10 pennies and put 1 dime in each compartment. If your child doesn't know this, then spend some time explaining the coins.

Parent: *Let's pretend we are at the store, and I want to buy a stuffed animal that costs 12 cents– but I only have a dime and 2 pennies in my pocket. Can I buy the stuffed animal? I will count the money in my pocket to see if I have enough. Remember, a dime is the same as 10 pennies. 10, 11, 12. I can buy the stuffed animal!*

Count 10, 11, 12 while pointing to the money on the table in front of you.

Parent: *Now we are going to write and build numbers. This is how to write the number 36.*

Write the number 36.

Parent: *Now I will show you how to build the number 36 by counting it out in dimes and pennies. The number 36 has 3 dimes and 6 pennies. I count the dimes by tens and the pennies by ones. Listen. 10, 20, 30, 31, 32, 33, 34, 35, 36.*

Place 3 dimes and 6 pennies on the table in front of you. Point to the dimes as you count 10, 20, 30. Point to the pennies as you count 31, 32, 33, 34, 35, 36.

Parent: *Now it is your turn to write and build numbers. Write the number 41.*

Child should write the number 41.

Parent: *Now build the number 41 by counting it out in dimes and pennies.*

Child should place 4 dimes and 1 penny on the table and count 10, 20, 30, 40, 41 while simultaneously pointing to each coin.

Parent: *How many dimes and how many pennies do you have? Leave the dimes and pennies next to where you wrote the number.*

Child should answer that he/she has 4 dimes and 1 penny.

Parent: *Write the number 28.*

Child should write the number 28.

Parent: *Now build the number 28 by counting it out in dimes and pennies.*

Child should place 2 dimes and 8 pennies on the table and count 10, 20, 21, 22, 23, 24, 25, 26, 27, 28 while simultaneously pointing to each coin.

Parent: *How many dimes and how many pennies do you have? Leave the dimes and pennies next to where you wrote the number.*

Child should answer that he/she has 2 dimes and 8 pennies.

Parent: *Look! You knew the answer to how many dimes and pennies you had before you built the number. You wrote it here and here. You are very clever!*

Point to where your child wrote the numbers 41 and 28.

Parent: *Look! You put 4 dimes where you wrote the number 4 and 1 penny where you wrote the number 1. Then you put 2 dimes and 8 pennies where you wrote 28. See how clever you are.*

Point again to where your child wrote the numbers 41 and 28.

Parent: *Write the number 56.*

Child should write the number 56.

Parent: *Now build the number 56 by counting it out in dimes and pennies.*

Child should place 5 dimes and 6 pennies on the table and count 10, 20, 30, 40, 50, 51, 52, 53, 54, 55, 56 while simultaneously pointing to each coin.

Parent: *How many dimes are in 56?*

Child should answer that there are 5 dimes in 56.

Parent: *Now I will use a new word. The word is tens. It means the same as dimes. How many tens are in the number 56? Think of how many dimes are in the number 56. I use the word tens because a dime is the same as 10 pennies.*

Child should answer that there are 5 tens in 56.

Parent: *Now I will use another new word. The word is ones. It means the same as pennies. How many ones are in the number 56? Think of how many pennies are in the number 56. Remember,*

I use the word pennies because a penny is the same as 1 cent.

Child should answer that there are 6 ones in 56.

Parent: *Now let's consider some other numbers. Write the number 43. Now build the number 43. Remember, to build a number means to count it out in dimes and pennies. How many dimes or tens are in 43? How many pennies or ones are in 43?*

Child should write 43, lay out 4 dimes and 3 pennies and answer that there are 4 tens and 3 ones in the number 43.

Parent: *Write and build the number 87. How many dimes or tens are in 87? How many pennies or ones are in 87?*

Child should write 87, lay out 8 dimes and 7 pennies and answer that there are 8 tens and 7 ones in the number 87.

Parent: *The digits in a number represent something called place value. A two digit number like the number 87 has a tens place and a ones place. In the number 87, the 8 is in the tens place and the 7 is in the ones place. By looking at the written number 87, we read that there are 8 tens and 7 ones.*

Parent: *I will write down some numbers. I want you to build the numbers with dimes and pennies and then tell me how many tens and how many ones. Try the number 24. What about 76?*

1. Child should write the number 24, lay out 2 dimes and 4 pennies and answer that there are 2 tens and 4 ones in the number 24.
2. Child should write the number 76, lay out 7 dimes and 6 pennies and answer that there are 7 tens and 6 ones in the number 76.

Parent: *Great! Now you understand place value. I wonder if you can trick me. Write down some 2-digit numbers. Let's see if I can build the numbers and tell you how many tens and ones.*

Be sure to have your child write down a number like 20 as one of his/her questions. Explain that 20 is 2 tens and 0 ones or no ones.

Lesson 34

In This Lesson You Will:

1. Count with your child to 120 while stepping/clapping
2. Count with your child to 120 while playing catch
3. Have your child place 10 objects in each of the 12 compartments
4. Model counting objects by 1 with "counting fingers"
5. Count objects together by 1 with "counting fingers"
6. Have your child count objects by 1 with "counting fingers"
7. Model skip counting by 10's
8. Model the "whisper/loud" counting trick
9. Count objects together using the "whisper/loud" counting trick
10. Have your child count objects using the "whisper/loud" counting trick
11. Have your child skip count by 2's, 3's, 4's, 5's, 6's, 7's, 8's and 9's on fingers and review
12. Have your child complete the workbook lesson
13. Recite or sing the skip counting chant together with any siblings or friends

Parent: *Now it is time to count our money. Put 10 pennies into each compartment.*

Your child may ask how he/she can put 10 pennies in each compartment when he/she only has 9 pennies, not 120 pennies.

Parent: *Do you see anything on the table that is the same as 10 pennies?*

If you have already done Lesson 33, your child may remember that a dime is the same as 10 pennies and put 1 dime in each compartment. If he/she does not know the value of a dime, spend some time explaining the coins.

Parent: *Now it is my turn to count the money.*

Lay out 10 pennies to help your child count each dime.

Parent: *Now it is my turn to count by 1's. Get ready. Count: 1, 2, 3...120.*

Lay the pennies out on the table. Point to each penny to count 1, 2, 3...9, then point to the first dime for 10. Repeat, pointing at the second dime for 20, and so on. Continue to 120.

Be sure to use your counting fingers.

Parent: *Now we'll count by 1's together. Are your counting fingers ready? Get ready. Count: 1, 2, 3...120.*

Have your child count along with you, using the pennies and dimes as described above.

Parent: *Now it is your turn to count by 1's. Get ready. Count.*

Child should count from 1 to 120, pointing with his/her counting fingers to each penny or dime as described above.

Parent: *Wonderful counting! Now we'll use our secret counting trick to count these dimes. When we point to the 1st penny, we will whisper 1. When we point to the 2nd second penny, we will whisper 2. When we point to the 3rd penny, we will whisper 3, and so on. When we point to the 10th penny, which is the dime, then we will say 10 out loud. Get ready. Count! 1, 2, 3...120.*

Point to the money while you explain the directions. Use whisper/loud technique.

Parent: *Now it is your turn to count this money. Use your counting fingers and the whisper/loud secret counting trick. (Shh). Get ready! Count!*

Child should count to 120 using the whisper/loud technique and his/her counting fingers. Continue to have your child practice this by counting dimes, or other groups of 10's.

Parent: *Excellent counting. Now it is my turn to count by 10's. Watch me count: 10, 20, 30, 40, 50, 60, 70, 80, 90, 100, 110, 120.*

Parent: *Now it is your turn to count by 10's.*

Child should count 10, 20, 30, 40, 50, 60, 70, 80, 90 100, 110, 120. Child will most likely enjoy counting by 10's more than any other number and be able to do it quite easily. Enjoy his/her enthusiasm! If your child needs more help counting by 10's, play My Turn, Your Turn to increase his/her speed.

Parent: *Now it is time to review. Wiggle your fingers and then skip count by 2's, 3's, 4's, 5's, 6's, 7's, 8's and 9's on your fingers.*

Child skip counts using his/her fingers. Use 2 of your fingers for the 11th and 12th fingers.

Parent: *What is 4×2?*

Child holds up 4 fingers and skip counts by 2's. Child should answer that 4 times 2 is 8.

Parent: *What is 4×3?*

Child holds up 4 fingers and skip counts by 3's. Child should answer that 4 times 3 is 12.

Parent: *What is 4×4?*

Child holds up 4 fingers and skip counts by 4's. Child should answer that 4 times 4 is 16.

Parent: *What is 4×5?*

Child holds up 4 fingers and skip counts by 5's. Child should answer that 4 times 5 is 20.

Parent: *What is 4×6?*

Child holds up 4 fingers and skip counts by 6's. Child should answer that 4 times 6 is 24.

Parent: *What is 4×7?*

Child holds up 4 fingers and skip counts by 7's. Child should answer that 4 times 7 is 28.

Parent: *What is 4×8?*

Child holds up 4 fingers and skip counts by 8's. Child should answer that 4 times 8 is 32.

Parent: *What is 4×9?*

Child holds up 4 fingers and skip counts by 9's. Child should answer that 4 times 9 is 36.

Lesson 35

In This Lesson You Will:

1. Count with your child to 120 while stepping/clapping
2. Count with your child to 120 while catching
3. Have your child place 10 objects in each of the 12 compartments
4. Model counting objects by 1 using the "whisper/loud" counting trick
5. Count objects together by 1 using the "whisper/loud" counting trick
6. Have your child count objects by 1 using the "whisper/loud" counting trick
7. Model skip counting by 10's
8. Skip count together by 10's
9. Play the Sleeping Bears game; substitute 10 into script on page 38
10. Have your child skip count by 10's
11. Play the Add 10 game (see below)
12. Play the Adding 10 game (see below)
13. Have your child skip counts by 2's, 3's, 4's, 5's, 6's, 7's, 8's and 9's on fingers and review
14. Have your child complete the workbook lesson
15. Recite or sing the skip counting chant together with any siblings and friends

Parent: *Now it is time to play a game called Add 10. When we add 10, we increase the tens place by 1. Let's use our dimes and pennies to help us play the game. Put 2 dimes and 3 pennies on the table. What is this number?*

Child should answer that it is 23 because there are 2 tens and 3 ones.

Parent: *It is my turn to add 10 to 23. When I add 10 to 23, I increase the 2 by 1 because the 2 is in the tens place. So my new number is 33.*

Parent: *Now add 10 to 33. What is the new number?*

Child should answer that when he adds 10 to 33, he increases the 3 by 1 because the 3 is in the tens place. So the new number is 43. Have child count dimes by tens to prove it!

Parent: *Put 8 dimes and 1 penny on the table. What is this* *number?*

Child should answer that it is 81 because there are 8 tens and 1 one.

Parent: *Now add 10 to 81. What is the new number?*

Child should answer that when he adds 10 to 81, he increases the 8 by 1 because the 8 is in the tens place. So the new number is 91.

Parent: *Put 4 more pennies on the table. What is this number?*

Child should answer that it is 95 because there are 9 tens and 5 ones.

Parent: *Now add 10 to 95. This one is tricky. 10 dimes is the same as 100 pennies. 100 is a 3 digit number that has a hundreds place as well as a tens and an ones place. The number 100 has a zero in the tens and ones place. What is the new number?*

Child should answer that when he adds 10 to 95, he increases the 9 by 1 because the 9 is in the tens place. 9 dimes plus 1 dime is 10 dimes, which is 1 dollar, or 100 pennies. This means that the new number is 105.

Parent: *Now we will play an even harder game called Adding 10. I will choose some dimes and pennies for our starting point. I choose 7 dimes and 4 pennies so our number is 74. Now hand me 1 dime at a time because I want to keep adding on 10. I will keep adding ten until we run out of dimes. 74 plus 10 is 84, 84 plus 10 is 94, 94 plus 10 is 104, 104 plus 10 is 114 and 114 plus 10 is 124.*

Parent: *It is your turn to add on tens. Let's start with 31. I will hand you the dimes one at a time. What are the new numbers?*

Give your child 3 dimes and 1 penny for the starting number 31. Child should answer that when he adds 10 to 31, he gets 41, then 51, 61, 71, 81, 91, 101, 111, 121.

Continue to play Add on 10 until your child is comfortable adding on 10 to any number.

Parent: *Let's review our multiplication. Wiggle your fingers and* *then skip count by 2's, on our fingers. (Follow with 3's, 4's, 5's, 6's, 7's, 8's and 9's.)*

Child skip counts using fingers. Use 2 of your fingers for the 11th and 12th fingers.

Parent: *What is 7×4?*

Child holds up 7 fingers and skip counts. Child should answer that 7 times 4 is 28.

Parent: *What is 7×5?*

Child holds up 7 fingers and skip counts. Child should answer that 7 times 5 is 35.

Parent: *What is 7×6?*

Child holds up 7 fingers and skip counts. Child should answer that 7 times 6 is 42.

Lesson 36

In This Lesson You Will:

1. Count with your child to 120 while stepping/clapping
2. Count with your child to 120 while catching
3. Have your child place 10 objects in each of the 12 compartments
4. Model counting objects by 1 using the "whisper/loud" counting trick
5. Count objects together by 1 using the "whisper/loud" counting trick
6. Have your child count objects by 1 using the "whisper/loud" counting trick
7. Model skip counting by 10's
8. Skip count together by 10's
9. Have your child skip count by 10's
10. Play the Add On 10 game; substitute 10 into script on page 39
11. Redo place value lesson on page 88 if necessary
12. Play Add 10, see script on page 96
13. Play the How Many? game, substitute 10 into script on page 43
14. Have your child skip count by 2's, 3's, 4's, 5's, 6's, 7's, 8's and 9's on fingers and review
15. Have your child complete the workbook lesson
16. Recite or sing the skip counting chant together with any siblings or friends

Lesson 37

In This Lesson You Will:

1. Count with your child to 120 while stepping/clapping
2. Count with your child to 120 while catching
3. Have your child place 10 objects in each of the 12 compartments
4. Model counting objects by 1 using the "whisper/loud" counting trick
5. Count objects together by 1 using the "whisper/loud" counting trick
6. Have your child count objects by 1 using the "whisper/loud" counting trick
7. Model skip counting by 10's
8. Skip count together by 10's
9. Have your child skip count by 10's
10. Redo place value lesson on page 88 if necessary
11. Play the How Many? game; substitute 10 into script on page 43
12. Play the Finger Counting game; substitute 10 into script on page 55
13. Model using fingers to solve multiplication problems
14. Practice solving multiplication problems together
15. Have your child solve multiplication problems
16. Have your child skip count by 2's, 3's, 4's, 5's, 6's, 7's, 8's and 9's on fingers and review
17. Look together for the fastest way to multiply
18. Have your child complete the workbook lesson
19. Recite or sing the skip counting chant together with any siblings or friends

Parent: *Let's review our multiplication. Wiggle your fingers and then skip count by 2's, on our fingers. (Follow with 3's, 4's, 5's, 6's, 7's, 8's and 9's.)*

Child skip counts using fingers. Borrow 2 fingers from you for the 11th and 12th finger.

Parent: *What's 8×3? In the past we have always held up 8 fingers and skip counted by 3's. What if we held up 3 fingers and skip counted by 8's? Would we get the same answer?*

Help your child to skip count on his/her fingers to discover that 8×3 and 3×8 are both equal to 24.

Parent: *Do you want to be fast when you multiply? Is it faster to skip count by the smaller number or the larger number?*

Child should answer that it is faster to hold up fingers for the smaller number and skip count by the larger number.

If your child is uncomfortable skip counting by the large numbers such as 7, 8 or 9, then he/she may prefer to skip count for longer, by a number that he/she is comfortable with, such as 2, 3 or 5. Continue to encourage practicing skip counting with the larger numbers in order to increase his/her speed in multiplication.

Parent: *What is 9×2? What is the fastest way to solve this?*

Child should answer that it is faster to hold up 2 fingers and skip count by 9's than it is to hold up 9 fingers and skip count by 2's. 9×2 and 2×9 are both equal to 18.

Parent: *What is 9×3? What is the fastest way to solve this?*

Child should answer that it is faster to hold up 3 fingers and skip count by 9's than it is to hold up 9 fingers and skip count by 3's. 9×3 and 3×9 are both equal to 27.

Parent: *What is 9×4? What is the fastest way to solve this?*

Child should answer that it is faster to hold up 4 fingers and skip count by 9's than it is to hold up 9 fingers and skip count by 4's. 9×4 and 4×9 are both equal to 36.

Parent: *What is 9×5? What is the fastest way to solve this?*

Child should answer that it is faster to hold up 5 fingers and skip count by 9's than it is to hold up 9 fingers and skip count by 5's. 9×5 and 5×9 are both equal to 45.

It may or may not be faster for your child to hold up 5 fingers and skip count by 9's. Counting by 5's is so easy for children that they may prefer this to a shorter skip count by 9's.

Parent: *What is 9×6? What is the fastest way to solve this?*

Child should answer that it is faster to hold up 6 fingers and skip count by 9's than it is to hold up 9 fingers and skip count by 6's. 9×6 and 6×9 are both equal to 54.

Parent: *What is 9×7? What is the fastest way to solve this?*

Child should answer that it is faster to hold up 7 fingers and skip count by 9's than it is to hold up 9 fingers and skip count by 7's. 9×7 and 7×9 are both equal to 63.

Lesson 38

In This Lesson You Will:

1. Count with your child to 132 while stepping/clapping
2. Count with your child to 132 while playing catch
3. Have your child place 11 objects in each of the 12 compartments
4. Model counting objects by 10's and 1's with "counting fingers"
5. Count objects together by 10's and 1's with "counting fingers"
6. Have your child count objects by 10's and 1's with "counting fingers"
7. Model skip counting by 11's
8. Model the "whisper/loud" counting trick
9. Count objects together using the "whisper/loud" counting trick
10. Have your child count objects using the "whisper/loud" counting trick
11. Have your child skip count by 2's, 3's, 4's, 5's, 6's, 7's, 8's, 9's and 10's on fingers and review
12. Have your child complete the workbook lesson
13. Recite or sing the skip counting chant together with any siblings or friends

Parent: *Now it is time to count our money. Put 11 pennies into each compartment.*

Your child may ask how he can put 10 pennies in each compartment when he/she only has 9 pennies, not 120 pennies.

Parent: *Do you see anything on the table that is the same as 11 pennies?*

Your child may remember that a dime is the same as 10 pennies and put 1 dime and one penny in each compartment. If your child doesn't remember this from Lesson 33, then spend some time explaining the coins.

Parent: *Now it is my turn to count the money.*

Lay out 10 pennies to help your child count each dime.

Parent: *Do you remember how to add on 10? My counting will be by 10's and 1's. 10, 11, 21, 22, 32, 33, 43, 44, 54, 55, 65, 66, 76, 77, 87, 88, 98, 99, 109, 110, 120, 121, 131, 132.*

Point to the dime and say 10, then point to the penny in the same compartment and say

11. Point to the dime in the next compartment and say 21, then point to the penny next to it and say 22. Point to the dimes each time you add on 10 and point to the pennies as you add on the 1's.

Parent: *Now we'll count by 10's and 1's together. Are your counting fingers ready? Get ready. Count: 10, 11, 21, 22, 32, 33, 43, 44, 54, 55, 65, 66, 76, 77, 87, 88, 98, 99, 109, 110, 120, 121, 131, 132.*

Have your child count along with you. Help him/her point to the correct dime or penny.

Parent: *It's your turn to count by 10's and 1's. Ready? Count.*

Child should count pointing to each penny or dime as described above. Be sure he/she uses his/her counting fingers.

Parent: *Excellent counting. Now it is my turn to skip count by 11's. Watch me count: 11, 22, 33, 44, 55, 66, 77, 88, 99, 110, 121, 132. Did you hear all the double digits? 11 is two ones. 22 is two twos. 33 is two threes. 44 is two fours. And so on, until 99 where 99 is two nines. Doesn't that make skip counting by 11's really easy?*

Lesson 39

In This Lesson You Will:

1. Count with your child to 132 while stepping/clapping
2. Count with your child to 132 while catching
3. Have your child place 11 objects in each of the 12 compartments
4. Model counting objects by 10's and 1's, see script on page 102
5. Count objects by 10's and 1's
6. Have your child count objects by 10's and 1's
7. Model skip counting by 11's
8. Skip count together by 11's
9. Play the Sleeping Bears game; substitute 11 into script on page 38
10. Have your child skip count by 11's
11. Play the Add On 11 game; substitute 11 into script on page 39
12. Have your child skip count by 2's, 3's, 4's, 5's, 6's, 7's, 8's, 9's and 10's on fingers and review
13. Have your child complete the workbook lesson
14. Recite or sing the skip counting chant together with any siblings or friends

Lesson 40

In This Lesson You Will:

1. Count with your child to 132 while stepping/clapping
2. Count with your child to 132 while catching
3. Have your child place 11 objects in each of the 12 compartments
4. Model counting objects by 10's and 1's
5. Count objects together by 10's and 1's
6. Have your child count objects by 10's and 1's
7. Model skip counting by 11's
8. Skip count together by 11's
9. Have your child skip count by 11's
10. Play the Add On 11 game; substitute 11 into script on page 39
11. Play the How Many? game; substitute 11 into script on page 43
12. Have the child skip count by 2's, 3's, 4's, 5's, 6's, 7's, 8's, 9's and 10's on fingers and review
13. Have your child complete the workbook lesson
14. Recite or sing the skip counting chant together with any siblings or friends

Lesson 41

In This Lesson You Will:

1. Count with your child to 132 while stepping/clapping
2. Count with your child to 132 while catching
3. Have your child place 11 objects in each of the 12 compartments
4. Model counting objects by 10's and 1's
5. Count objects together by 10's and 1's
6. Have your child count objects by 10's and 1's
7. Model skip counting by 11's
8. Skip count together by 11's
9. Have your child skip count by 11's
10. Play the How Many? game; substitute 11 into script on page 43
11. Play the Finger Counting game; substitute 11 into script on page 55
12. Model using fingers to solve multiplication problems
13. Practice solving multiplication problems together
14. Have your child solve multiplication problems
15. Have your child skip count by 2's, 3's, 4's, 5's, 6's, 7's, 8's, 9's and 10's on fingers and review
16. Have your child complete the workbook lesson
17. Recite or sing the skip counting chant together with siblings or friends

Lesson 42

In This Lesson You Will:

1. Count with your child to 144 while stepping/clapping
2. Count with your child to 144 while playing catch
3. Have your child place 12 objects in each of the 12 compartments
4. Model counting objects by 10's and 1's with "counting fingers"(see below)
5. Count objects together by 10's and 1's with "counting fingers"
6. Have your child count objects by 10's and 1's with "counting fingers"
7. Model skip counting by 12's
8. Model the "whisper/loud" counting trick
9. Count objects together using the "whisper/loud" counting trick
10. Have your child count objects using the "whisper/loud" counting trick
11. Have your child skip count by 2's, 3's, 4's, 5's, 6's, 7's, 8's, 9's, 10's and 11's on fingers and review
12. Have your child complete the workbook lesson
13. Recite or sing the skip counting chant together with any siblings or friends

Parent: *Now it is time to count our money. Put 12 cents into each compartment.*

Your child may ask how he/she can put 12 pennies in each compartment when he/she only has 12 pennies, not 144 pennies.

Parent: *Do you see anything on the table that is the same as 12 pennies?*

Your child should remember that a dime is the same as 10 pennies and put 1 dime and 2 pennies in each compartment. If your child doesn't remember this from previous lessons, then spend some time explaining the coins.

Parent: *Now it is my turn to count the money.*

Lay out 10 pennies to help your child count each dime.

Parent: *Do you remember how to add on 10? My counting will be by 10's and 1's. 10, 11, 12, 22, 23, 24, 34, 35, 36, 46, 47, 48, 58, 59, 60, 70, 71, 72, 82, 83, 84, 94, 95, 96, 106, 107, 108, 118, 119, 120, 130, 131, 132, 142, 143, 144.*

Point to the dime and say 10, then point to the penny in the same compartment, and say 11. Point to the other penny in the same compartment, and say 12. Point to the dime in the next compartment and say 22, then point to the penny in the same compartment and say 23. Point to the other penny in the same compartment, and say 24. Point to the dimes each time you add on 10 and point to the pennies as you add on the 1's.

Parent: *Now we'll count by 10's and 1's together. Are your counting fingers ready? Get ready. Count: 10, 11, 12, 22, 23, 24, 34, 35, 36, 46, 47, 48, 58, 59, 60, 70, 71, 72, 82, 83, 84, 94, 95, 96, 106, 107, 108, 118, 119, 120, 130, 131, 132, 142, 143, 144.*

Have your child count along with you. Help him/her point to the correct dime or penny.

Parent: *It's your turn to count by 10's and 1's. Ready? Count.*

Child should count pointing to each penny or dime as described above. Be sure he/she uses his/her counting fingers.

Lesson 43

In This Lesson You Will:

1. Count with your child to 144 while stepping/clapping
2. Count with your child to 144 while catching
3. Have your child place 12 objects in each of the 12 compartments
4. Model counting objects by 10's and 1's (see script on page 107)
5. Count objects together by 10's and 1's
6. Have your child count objects by 10's and 1's
7. Model skip counting by 12's
8. Skip count together by 12's
9. Demonstrate that your child already knows the skip counting by 12's by having your child skip count by 3's and 4's.
10. Play My Turn, Your Turn
11. Play the Sleeping Bears game; substitute 12 into script on page 38
12. Have your child skip count by 12's
13. Play the Add On 12 game; substitute 12 into script on page 39
14. Have your child skip count by 2's, 3's, 4's, 5's, 6's, 7's, 8's, 9's, 10's and 11's on fingers and review
15. Have your child complete the workbook lesson
16. Recite or sing the skip counting chant together with any siblings or friends

Parent: *Excellent counting. Now it is my turn to skip count by 12's. Here are the first 5. 12, 24, 36, 48, 60. Do you remember that when we skip counted by 2's we counted to 24?*

Have your child skip count by 2's from 2 to 24 to help him/her see that 24 is in the 2 and 12's skip counting pattern.

Parent: *Do you remember that when we skip counted by 3's we counted to 36?*

Have your child skip count by 3's from 3 to 36 to help him/her see that 36 is in the 3 and 12's skip counting pattern.

Parent: *Do you remember that when we skip counted by 4's we counted to 48?*

Have your child skip count by 4's from 4 to 48 to help him/her see that 48 is in the 4 and 12's skip counting pattern.

Parent: *Skip counting by 12's will be easier to remember if you think of the last number in the skip counting pattern when we skip counted by 2's, 3's 4's, 5's, 6's, 7's, 8's, 9's, 10's, and 11's.*

Parent: *Now it is my turn again to skip count by 12's. 12, 24, 36, 48, 60, 72, 84, 96, 108, 120, 132, 144.*

Parent: *Now it is your turn to skip count by 12's. Touch each compartment as you count.*

Child should count 12, 24, 36, 48, 60, 72, 84, 96, 108, 120, 132, 144, touching the dime and 2 pennies as he/she counts. Make sure he/she uses his/her counting fingers.

Parent: *Let's play My Turn, Your Turn. Say 12.*

Child should say 12.

Parent: *12, 24. Now say 12, 24, 36.*

Child should say 12, 24, 36.

Parent: *12, 24, 36, 48. Now say 12, 24, 36, 48, 60.*

Child should say 12, 24, 36, 48, 60.

Parent: *12, 24, 36, 48, 60, 72. Now say 12, 24, 36, 48, 60, 72, 84.*

Child should say 12, 24, 36, 48, 60, 72, 84.

Parent: *12, 24, 36, 48, 60, 72, 84, 96. Now say 12, 24, 36, 48, 60, 72, 84, 96, 108.*

Child should say 12, 24, 36, 48, 60, 72, 84, 96, 108.

Parent: *12, 24, 36, 48, 60, 72, 84, 96, 108, 120. Now say 12, 24, 36, 48, 60, 72 84, 96, 108, 120, 132.*

Child should say 12, 24, 36, 48, 60, 72, 84, 96, 108, 120, 132.

Parent: *Now we'll skip count by 12's together. 12, 24, 36, 48, 60, 72 84, 96, 108, 120, 132, 144.*

Parent: *Now it is your turn to skip count by 12's.*

Child should skip count 12, 24, 36, 48, 60, 72 84, 96, 108, 120, 132, 144.

Lesson 44

In This Lesson You Will:

1. Count with your child to 144 while stepping/clapping
2. Count with your child to 144 while catching
3. Have your child place 12 objects in each of the 12 compartments
4. Model counting objects by 10's and 1's (see script on page 107)
5. Count objects together by 10's and 1's
6. Have your child count objects by 10's and 1's
7. Model skip counting by 12's
8. Skip count together by 12's
9. Play My Turn, Your Turn; see script on page 110
10. Have your child skip count by 12's
11. Play the Add On 12 game; substitute 12 into script on page 39
12. Play the How Many? game; substitute 12 into script on page 43
13. Have your child skip count by 2's, 3's, 4's, 5's, 6's, 7's, 8's, 9's, 10's and 11's on fingers and review
14. Have your child complete the workbook lesson
15. Recite or sing the skip counting chant together with any siblings or friends

Lesson 45

In This Lesson You Will:

1. Count with your child to 144 while stepping/clapping
2. Count with your child to 144 while catching
3. Have your child place 12 objects in each of the 12 compartments
4. Model counting objects by 10's and 1's (see script on page 107)
5. Count objects together by 10's and 1's
6. Have your child count objects by 10's and 1's
7. Model skip counting by 12's
8. Skip count together by 12's
9. Play My Turn, Your Turn; see script on page 110
10. Have your child skip count by 12's
11. Play the How Many? game; substitute 12 into script on page 43
12. Play the Finger Counting game; substitute 12 into script on page 55
13. Model using fingers to solve multiplication problems
14. Practice solving multiplication problems together
15. Have your child solve multiplication problems
16. Have your child skip count by 2's, 3's, 4's, 5's, 6's, 7's, 8's, 9's, 10's and 11's on fingers and review
17. Have your child complete the workbook lesson
18. Recite or sing the skip counting chant together with any siblings or friends

SKIP COUNTING CHANT

In each lesson, the last activity is to sing or recite the skip counting chant. The skip counting chant is skip counting by 2's, 3's, 4's, 5's, 6's, 7's, 8's, 9's, 10's, 11's, and 12's. It begins on page 114 and concludes on page 122.

Have your child color in each skip counting pattern found on pages 114 to122, according to the directions. For ease of use, you may prefer to copy, color in, and laminate pages 114 to122.

Directions: Have your child color in 2, 4, 6, 8, 10, 12, 14, 16, 18, 20, 22, and 24 using a light-colored crayon.

Skip Counting by Twos

1	2	3	4	5	6	7	8	9	10
11	12	13	14	15	16	17	18	19	20
21	22	23	24						

Skip Counting by Threes

Directions: Have your child color in 3, 6, 9, 12, 15, 18, 21, 24, 27, 30, 33, and 36 using a light-colored crayon.

1	2	3	4	5	6	7	8	9	10
11	12	13	14	15	16	17	18	19	20
21	22	23	24	25	26	27	28	29	30
31	32	33	34	35	36				

Skip Counting by Fours

Directions: Have your child color in 4, 8, 12, 16, 20, 24, 28, 32, 36, 40, 44, and 48 using a light-colored crayon.

1	2	3	4	5	6	7	8	9	10
11	12	13	14	15	16	17	18	19	20
21	22	23	24	25	26	27	28	29	30
31	32	33	34	35	36	37	38	39	40
41	42	43	44	45	46	47	48		

Skip Counting by Fives

Directions: Have your child color in 5, 10, 15, 20, 25, 30, 35, 40, 45, 50, 55, and 60 using a light-colored crayon.

1	2	3	4	5	6	7	8	9	10
11	12	13	14	15	16	17	18	19	20
21	22	23	24	25	26	27	28	29	30
31	32	33	34	35	36	37	38	39	40
41	42	43	44	45	46	47	48	49	50
51	52	53	54	55	56	57	58	59	60

Skip Counting by Sixes

Directions: Have your child color in 6, 12, 18, 24, 30, 36, 42, 48, 54, 60, 66, and 72 using a light-colored crayon.

1	2	3	4	5	6	7	8	9	10
11	12	13	14	15	16	17	18	19	20
21	22	23	24	25	26	27	28	29	30
31	32	33	34	35	36	37	38	39	40
41	42	43	44	45	46	47	48	49	50
51	52	53	54	55	56	57	58	59	60
61	62	63	64	65	66	67	68	69	70
71	72								

Skip Counting by Sevens

Directions: Have your child color in 7, 14, 21, 28, 35, 42, 49, 56, 63, 70, 77, and 84 using a light-colored crayon.

1	2	3	4	5	6	7	8	9	10
11	12	13	14	15	16	17	18	19	20
21	22	23	24	25	26	27	28	29	30
31	32	33	34	35	36	37	38	39	40
41	42	43	44	45	46	47	48	49	50
51	52	53	54	55	56	57	58	59	60
61	62	63	64	65	66	67	68	69	70
71	72	73	74	75	76	77	78	79	80
81	82	83	84						

Skip Counting by Eights

Directions: Have your child color in 8, 16, 24, 32, 40, 48, 56, 64, 72, 80, 88, and 96 using a light-colored crayon.

1	2	3	4	5	6	7	8	9	10
11	12	13	14	15	16	17	18	19	20
21	22	23	24	25	26	27	28	29	30
31	32	33	34	35	36	37	38	39	40
41	42	43	44	45	46	47	48	49	50
51	52	53	54	55	56	57	58	59	60
61	62	63	64	65	66	67	68	69	70
71	72	73	74	75	76	77	78	79	80
81	82	83	84	85	86	87	88	89	90
91	92	93	94	95	96				

Skip Counting by Nines

Directions: Have your child color in 9, 18, 27, 36, 45, 54, 63, 72, 81, 90, 99, and 108 using a light-colored crayon.

1	2	3	4	5	6	7	8	9	10
11	12	13	14	15	16	17	18	19	20
21	22	23	24	25	26	27	28	29	30
31	32	33	34	35	36	37	38	39	40
41	42	43	44	45	46	47	48	49	50
51	52	53	54	55	56	57	58	59	60
61	62	63	64	65	66	67	68	69	70
71	72	73	74	75	76	77	78	79	80
81	82	83	84	85	86	87	88	89	90
91	92	93	94	95	96	97	98	99	100
101	102	103	104	105	106	107	108		

Skip Counting by Tens

Directions: Have your child color in 10, 20, 30, 40, 50, 60, 70, 80, 90, 100, 110, and 120 using a light-colored crayon.

1	2	3	4	5	6	7	8	9	10
11	12	13	14	15	16	17	18	19	20
21	22	23	24	25	26	27	28	29	30
31	32	33	34	35	36	37	38	39	40
41	42	43	44	45	46	47	48	49	50
51	52	53	54	55	56	57	58	59	60
61	62	63	64	65	66	67	68	69	70
71	72	73	74	75	76	77	78	79	80
81	82	83	84	85	86	87	88	89	90
91	92	93	94	95	96	97	98	99	100
101	102	103	104	105	106	107	108	109	110
111	112	113	114	115	116	117	118	119	120

Skip Counting by Elevens

Directions: Have your child color in 11, 22, 33, 44, 55, 66, 77, 88, 99, 110, 121, and 132 using a light-colored crayon.

1	2	3	4	5	6	7	8	9	10
11	12	13	14	15	16	17	18	19	20
21	22	23	24	25	26	27	28	29	30
31	32	33	34	35	36	37	38	39	40
41	42	43	44	45	46	47	48	49	50
51	52	53	54	55	56	57	58	59	60
61	62	63	64	65	66	67	68	69	70
71	72	73	74	75	76	77	78	79	80
81	82	83	84	85	86	87	88	89	90
91	92	93	94	95	96	97	98	99	100
101	102	103	104	105	106	107	108	109	110
111	112	113	114	115	116	117	118	119	120
121	122	123	124	125	126	127	128	129	130
131	132								

Skip Counting by Twelves

Directions: Have your child color in 12, 24, 36, 48, 60, 72, 84, 96, 108, 120, 132, and 144 using a light-colored crayon.

1	2	3	4	5	6	7	8	9	10
11	12	13	14	15	16	17	18	19	20
21	22	23	24	25	26	27	28	29	30
31	32	33	34	35	36	37	38	39	40
41	42	43	44	45	46	47	48	49	50
51	52	53	54	55	56	57	58	59	60
61	62	63	64	65	66	67	68	69	70
71	72	73	74	75	76	77	78	79	80
81	82	83	84	85	86	87	88	89	90
91	92	93	94	95	96	97	98	99	100
101	102	103	104	105	106	107	108	109	110
111	112	113	114	115	116	117	118	119	120
121	122	123	124	125	126	127	128	129	130
131	132	133	134	135	136	137	138	139	140
141	142	143	144						